RELATIVITY

D0815579

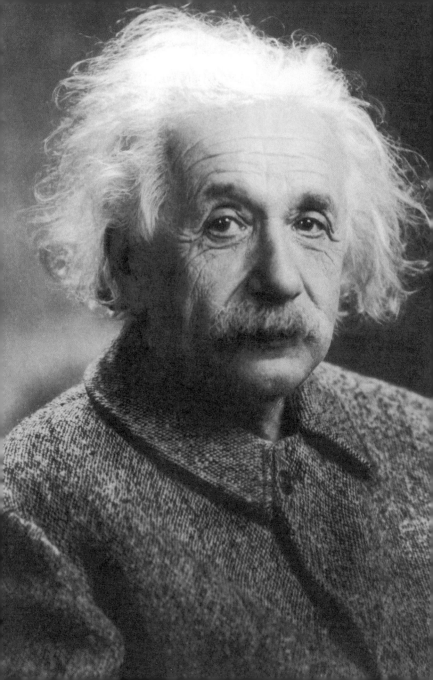

RELATIVITY

•

Russell Stannard

A BRIEF
INSIGHT

STERLING

New York / London
www.sterlingpublishing.com

STERLING and the distinctive Sterling logo are registered trademarks of
Sterling Publishing Co., Inc.

Library of Congress Cataloging-in-Publication Data

Stannard, Russell.
   Relativity / Russell Stannard. -- Illustrated ed.
       p. cm. -- (A brief insight)
   Includes index.
   ISBN 978-1-4027-7899-5
   1. Relativity (Physics) I. Title.
   QC173.55.S73 2010
   530.11--dc22

                              2010013455

10 9 8 7 6 5 4 3 2 1

Published by Sterling Publishing Co., Inc.
387 Park Avenue South, New York, NY 10016

Published by arrangement with Oxford University Press, Inc.

© 2008 by Russell Stannard
Illustrated edition published in 2011 by Sterling Publishing Co., Inc.
Additional text © 2011 Sterling Publishing Co., Inc.

Distributed in Canada by Sterling Publishing
c/o Canadian Manda Group, 165 Dufferin Street
Toronto, Ontario, Canada M6K 3H6

Book design: Faceout Studio

Please see picture credits on page 149 for image copyright information.

Printed in China
All rights reserved

Sterling ISBN 978-1-4027-7899-5

For information about custom editions, special sales, premium and corporate purchases, please contact
Sterling Special Sales Department at 800-805-5489 or specialsales@sterlingpublishing.com.

Frontispiece: Albert Einstein (1879–1955), whose theories of special relativity and general relativity
revolutionized the study of science, devised his famous equation $E = mc^2$—a formulation of the principle
of mass-energy equivalence—in 1905. This photograph of Einstein was taken around 1939.

To my grandchildren

Like myself, may they find the
exploration of Einstein's
thoughts a constant source of
wonder and fascination

•

CONTENTS

•

Preface VIII

ONE   Special Relativity I

The Principle of Relativity and the Speed of Light I

Time Dilation 6

The Twin Paradox 14

Length Contraction 17

Loss of Simultaneity 22

Space-Time Diagrams 26

Four-Dimensional Spacetime 32

The Ultimate Speed 41

$E = mc^2$ 45

TWO General Relativity 57

The Equivalence Principle 57

The Effects on Time of Acceleration and Gravity 64

The Twin Paradox Revisited 72

The Bending of Light. 78

Curved Space 84

Black Holes 99

Gravitational Waves 120

The Universe 126

Further Reading. 140

Index. 142

Picture Credits. 148

PREFACE

•

All of us grow up with certain basic ideas concerning space, time, and matter. These include:

We all inhabit the same three-dimensional space;
Time passes equally quickly for everyone;
Two events occur either simultaneously, or one before the other;
Given enough power, there is no limit to how fast one can travel;
Matter can be neither created nor destroyed;
The angles of a triangle add up to 180°;
The circumference of a circle is $2\pi \times$ the radius;
In a vacuum, light always travels in straight lines.

Such notions appear to be little more than common sense. But be warned:

Common sense consists of those layers of prejudice laid down in the mind before the age of eighteen.

Albert Einstein

In fact, Einstein's theory of relativity challenges all the above statements. There are circumstances in which each of them can be shown to be false. Startling as such findings are, it is not difficult to retrace Einstein's thinking. In this book we shall see how, starting from well-known everyday observations, coupled with the results of certain experiments, we can logically work our way to these conclusions. From time to time a little mathematics will be introduced, but nothing beyond the use of square roots and Pythagoras's theorem. Readers able and wishing to follow up with a more detailed mathematical treatment are referred to the further reading list.

The theory is divided into two parts: the *special theory of relativity*, formulated in 1905, and the *general theory of relativity*, which appeared in 1916. The former deals with the effects on space and time of uniform motion. The latter includes the additional effects of acceleration and of gravity. The former is a special case of the all-embracing general theory. It is with this special case that we begin . . .

ONE

Special Relativity

•

The Principle of Relativity and the Speed of Light

Imagine you are in a train carriage waiting at a station. Out of the window you see a second train standing alongside yours. The whistle blows, and at last you are on your way. You glide smoothly past the other train. Its last carriage disappears from view, allowing you to see the station also disappearing into the distance as it is left behind. Except that the station is *not* disappearing; it is just sitting there going nowhere—just as you are sitting in the train going nowhere. It dawns on you that you weren't moving at all; it was the *other* train which moved off.

A simple observation. We all get fooled this way at some time or other. The truth is that you cannot tell whether you are really on the

Whether one is sitting on a train in steady motion or a train at rest, everything that happens inside the enclosure of that train happens within an inertial frame of reference. If two trains are positioned side by side, their relative motion can be disorienting to observers in the trains themselves.

move or not—at least, not if we are talking about steady uniform motion in a straight line. Normally, when traveling by car, say, you do know that you are moving. Even if you have your eyes shut, you can feel pushed around as the car goes around corners, goes over bumps, speeds up or slows down suddenly. But in an aircraft cruising steadily, apart from the engine noise and the slight vibrations, you would have no way of telling that you were moving. Life carries on inside the plane exactly as it would if it were stationary on the ground. We say the plane provides an *inertial frame of reference*. By this we mean Newton's law of inertia applies; namely, when viewed from this reference frame, an object will neither change its speed nor direction unless acted upon by an unbalanced force. A glass of water on the tray table in front of you, for example, remains stationary until you move it with your hand.

But what if you look out of the aircraft window and see the earth passing by underneath? Does that not tell you that the plane is moving? Not really. After all, the earth is not stationary; it is moving in orbit about the sun; the sun itself is orbiting the center of the Milky Way Galaxy; and the Milky Way Galaxy is moving about within a cluster of similar galaxies. All we can say is that these movements are all *relative*. The plane moves relative to the earth; the earth moves relative to the plane. There is no way of deciding who is *really* stationary. Anyone moving uniformly with respect to another at rest is entitled to consider himself to be at rest and the other person moving. This is because the laws of nature—the rules governing all that goes on—are the same for everyone in uniform steady motion, that is to say, everyone in an inertial frame of reference. This is *the principle of relativity*.

And no, it was not Einstein who discovered this principle; it goes back to Galileo. That being so, why has the word "relativity" become associated with Einstein's name? What Einstein noticed was that among the laws of nature were Maxwell's laws of electromagnetism. According to Maxwell, light is a form of electromagnetic radiation. As such, it becomes possible, from a knowledge of the strengths of electric and magnetic forces, to calculate the speed of light, c, in a vacuum. The fact that light has a

In addition to his pioneering work in the field of electromagnetism, the Scottish physicist James Clerk Maxwell (1831–79) conducted important research into the motion of molecules in a gas. In 1859, he postulated that Saturn's rings comprise numerous independently moving particles—a theory that was confirmed in observations made by the Voyager space probes in the 1980s.

speed is not immediately obvious. When you go into a darkened room and switch on a lamp, the light appears to be everywhere—ceiling, walls, and floor—instantly. But it is not so. It takes time for the light to travel from the light bulb to its destination. Not much time—it's too fast to see the delay with the naked eye. According to this law of nature, the speed of light in a vacuum, c, works out to be 299,792.458 kilometers per second (or very slightly different in air). And that's what the speed is measured to be.

What if the source of light is moving? One might, for example, expect light to behave like a shell being fired from a passing warship where an observer on the seashore would expect the speed of the ship to be added to the shell's muzzle speed if being fired in the forward direction, and subtracted if being fired to the rear. The behavior of light in this regard was checked at the CERN laboratory in Geneva in 1964, using tiny subatomic particles called *neutral pions*. The pions, traveling at $0.99975c$, decayed with the emission of two light pulses. Both pulses were found to have the usual speed of light, c, to within the measurement accuracy of 0.1%. So, the speed of light does not depend on the speed of the source.

It also does not depend on whether the observer of the light is considered to be moving or not. Take the case of a moving vessel again. Having already established that light does not behave like a shell being fired from a gun, we might expect it to behave like the ripples on the water. If the observer were now someone aboard a moving boat, the wave front would appear to move ahead of the boat more slowly than the wave front going to the rear—because of the motion of the boat and of himself relative to the water (see figure 1). If light were a wave moving through a medium pervading all of space—a medium provisionally called the ether—then, with the earth ploughing its way through the ether, we ought to find the speed of light relative to us observers traveling along with the earth to be different in different directions. But in a famous experiment carried out by Michelson and Morley in 1887, the speed of light was found to be the same in all directions. Thus, the speed of light is independent of whether either the source or the observer is considered to be moving.

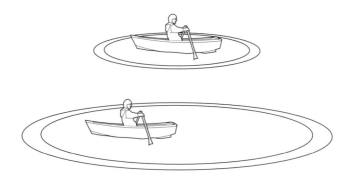

1. Ripples sent out by a boat appear to an observer on the boat to move away more slowly in the forward direction than to the rear.

So there we have it:

(i) The principle of relativity, which states that the laws of nature are the same for all inertial frames of reference.

(ii) One of those laws allows us to work out the value of the speed of light in a vacuum—a value which is the same in all inertial frames, regardless of the velocity of the source or the observer.

These two statements came to be known as the two *postulates* (or fundamental principles) of special relativity.

These facts had been common knowledge among physicists for a long time. It required the genius of Einstein to spot that although each of the two statements made sense when you thought about them separately, they did not appear to make sense if you put the two ideas together. It seemed as though if the first of them was right, then the second must be wrong, or if the second was right, the first must be wrong. If both were right—which we appear to have established—then something very, very serious must be amiss. The fact that the speed of light is the same for all inertial observers regardless of the motion of the source or observer means that our usual way of adding and subtracting velocities is wrong. And if there is something wrong with our conception of velocity (which is simply distance divided by time), then that in turn implies there must be something wrong with our conception of space, or time, or both. What we are dealing with is not some peculiarity of light or electromagnetic radiation. *Anything* traveling at the same speed as that of light will have the same value for its speed for all inertial observers. What is crucial is the speed (and the implications for the underlying space and time)—not the fact that we happen to be dealing with light.

Time Dilation

To see what is amiss, imagine an astronaut in a high-speed spacecraft and a mission controller on the ground. They both have identical clocks. The astronaut is to carry out a simple experiment. On the floor of the craft she is to fix a lamp which emits a pulse of light. The pulse travels directly upward at right angles to the direction of motion of the craft (see figure 2). There the pulse strikes a bull's-eye target fixed to the ceiling. Let us say that the height of the craft is 4 meters. With the light traveling at speed, c, she finds that the time taken for this trip, t', as measured on her clock, is given by $t' = 4/c$.

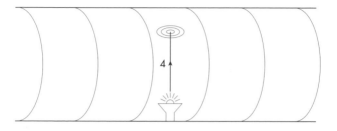

2. The astronaut arranges for a pulse of light to be directed toward a target such that the light travels at right angles to the direction of motion of the spacecraft.

Now let's see what this looks like from the perspective of the mission controller. As the craft passes him overhead, he too observes the trip performed by the light pulse from the source to the target. According to his perspective, during the time taken for the pulse to arrive at the target, the target will have moved forward from where it was when the pulse was emitted. For him, the path is not vertical; it slopes (see figure 3). The length of this sloping path will clearly be longer than it was from the astronaut's point of view. Let us say that the craft moves forward 3 meters in the time that it takes for the light pulse to travel from the source to the target. Using Pythagoras's theorem, where $3^2 + 4^2 = 5^2$, we see that the distance traveled by the pulse to get to the target is, according to the controller, 5 meters.

So what does he find for the time taken for the pulse to perform the trip? Clearly it is the distance traveled, 5 meters, divided by the speed at

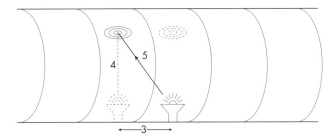

3. According to the mission controller on earth, as the spacecraft passes overhead, the target moves forward in the time it takes for the light pulse to perform its journey. The pulse, therefore, has to traverse a diagonal path.

which he sees the light traveling. This we have established is c (the same as it was for the astronaut). Thus, for the controller, the time taken, t, registered on his clock, is given by $t = 5/c$.

But this is not the time the astronaut found. She measured the time to be $t' = 4/c$. So, they disagree as to how long it took the pulse to perform the trip. According to the controller, the reading on the astronaut's clock is too low; her clock is going slower than his.

And it is not just the clock. Everything going on in the spacecraft is slowed down in the same ratio. If this were not so, the astronaut would be able to note that her clock was going slow (compared, say, to her heartbeat rate, or the time taken to boil a kettle, etc.). And that in turn would allow her to deduce that she was moving—her speed somehow affecting the mechanism of the clock. But that is not allowed by the principle of relativity. All uniform motion is relative. Life for the astronaut must proceed in exactly the same way as it does for the mission controller.

Thus we conclude that everything happening in the spacecraft—the clock, the workings of the electronics, the astronaut's aging processes, her thinking processes—all are slowed down in the same ratio. When she observes her slow clock with her slow brain, nothing will seem amiss. Indeed, as far as she is concerned, everything inside the craft keeps in step and appears normal. It is only according to the controller that everything in the craft is slowed down. This is *time dilation*. The astronaut has her time; the controller has his. They are not the same.

In that example we took a specific case, one in which the astronaut and spacecraft travel 3 meters in the time it takes light to travel the 5 meters from the source to the target. In other words, the craft is traveling at a speed of $^3/_5c$; i.e., $0.67c$. And for that particular speed we found that the astronaut's time was slowed down by a factor $^4/_5$; i.e., 0.8. It is easy enough to obtain a formula for any chosen speed, v. We apply Pythagoras's theorem to triangle ABC. The distances are as shown in figure 4. Thus:

$$(1) \qquad AC^2 = AB^2 + BC^2$$
$$AB^2 = AC^2 - BC^2$$
$$c^2t'^2 = (c^2 - v^2)t^2$$
$$t'^2 = (1 - v^2/c^2)t^2$$
$$t' = t\sqrt{(1 - v^2/c^2)}$$

From this formula we see that if v is small compared to c, the expression under the square root sign approximates to one, and $t' \approx t$. Yet no matter how small v becomes, the dilation effect is still there. This means that, strictly speaking, whenever we undertake a journey—say, a bus trip—on alighting we ought to readjust our watch to get it back into

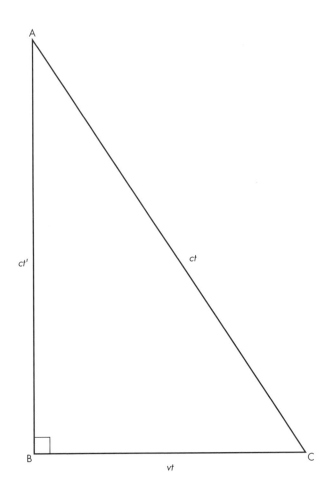

4. According to the mission controller, BC is the distance traveled by the craft in the time taken for the light pulse to travel to the target, and AC is the distance traveled by the pulse. AB is the distance traveled by the pulse according to the astronaut.

synchronization with all the stationary clocks and watches. The reason we do not is that the effect is so small. For instance, someone opting to drive express trains all their working life will get out of step with those following sedentary jobs by no more than about one-millionth of a second by the time they retire. Hardly worth bothering about.

At the other extreme, we see from the formula that, as v approaches c, the expression under the square root sign approaches zero, and t' tends to zero. In other words, time for the astronaut would effectively come to a standstill. This implies that if astronauts were capable of flying very close to the speed of light, they would hardly age at all and would, in effect, live for ever. The downside, of course, is that their brains would have almost come to a standstill, which in turn means they would be unaware of having discovered the secret of eternal youth.

So much for the theory. But is it true in practice? Emphatically, yes. In 1977, for instance, an experiment was carried out at the CERN laboratory in Geneva on subatomic particles called *muons*. These tiny particles are unstable, and after an average time of 2.2×10^{-6} seconds (i.e., 2.2 millionths of a second) they break up into smaller particles. They were made to travel repeatedly around a circular trajectory of about 14 meters diameter, at a speed of $v = 0.9994c$. The average lifetime of these moving muons was measured to be 29.3 times longer than that of stationary muons—exactly the result expected from the formula we have derived, to an experimental accuracy of 1 part in 2000.

In a separate experiment carried out in 1971, the formula was checked out at aircraft speeds using identical atomic clocks, one carried in an aircraft, and the other on the ground. Again, good agreement with theory was found. These and innumerable other experiments all confirm the correctness of the time dilation formula.

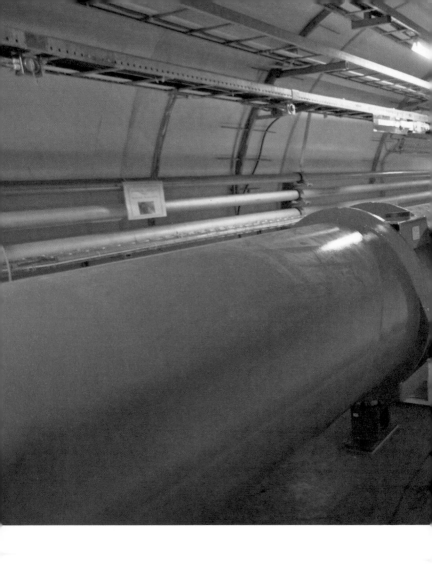

High-energy subatomic particles are accelerated and stored in giant circular machines, kept on course by magnets. It was in such a machine that unstable muons were stored as their lifetimes were measured. This photograph shows a small section of the largest and most powerful of these machines: the Large Hadron Collider at CERN (formerly the Conseil Européen pour la Recherche Nucléaire, now the European Organization for Nuclear Research).

The Twin Paradox

We have seen how the mission controller sees time passing slowly in the moving spacecraft, while the astronaut regards her time as normal. How does the *astronaut* see the *mission controller's* time?

At first one might think that if her time is going slow, then when she observes what is happening on the ground, she will perceive time down there to be going fast. But wait. That cannot be right. If it were, then we would immediately be able to conclude who was actually moving and who was stationary. We would have established that the astronaut was the moving observer because her time was affected by the motion whereas the controller's wasn't. But that violates the principle of relativity, i.e., that for inertial frames, all motion is relative. Thus, the principle leads us to the admittedly somewhat uncomfortable conclusion that if the controller concludes that the astronaut's clock is going slower than his, then she will conclude that his clock is going slower than hers. But how, you might ask, is that possible? How can we have two clocks, both of which are lagging behind the other?!

A preliminary to addressing this problem is that we must first recognize that in the set-up we have described we are not comparing clocks directly side-by-side. Though the astronaut and controller might indeed have synchronized their two clocks as they were momentarily alongside each other at the start of the space trip, they cannot do the same for the subsequent reading; the spacecraft and its clock have flown off into the distance. The controller can only find out how the astronaut's clock is doing by waiting for some kind of signal (perhaps a light signal) to be emitted by her clock and received by himself. He then has to allow for the fact that it has taken time for that signal to travel from the craft's new location to himself at mission control. By adding that transmission time to the reading of the clock when it emitted the signal, he can then calculate what the time is on the other clock now, and compare it with the reading on his own. It is only then that he concludes that the astronaut's clock is running slow. But note this is the result of a *calculation*, not a direct visual comparison. And the same will be true for the astronaut. She arrives at her conclusion that it is the controller's clock that is running slow only on the basis of a calculation using a signal emitted by his clock.

Which doubtless still leaves a nagging question in your mind, namely "But whose clock is *really* going slow?" With the set-up we have described, that is a meaningless question. It has no answer. As far as the mission controller is concerned, it is true that the astronaut's clock is the one going slow; as far as the astronaut is concerned, it is true that it is the mission controller's clock that is going slow. And we have to leave it at that.

Not that people have left it at that. Enter the famous *twin paradox*. This proposal recognizes that the seemingly contradictory conclusions arise because the times are being *calculated*. But what if the calculations could be replaced by direct side-by-side comparisons of the two clocks—at the end of the journey as well as at the beginning? That way there would be no ambiguity. What this would require is that the spacecraft, having traveled to, say, a distant planet, turns around and comes back home so that the two clocks can be compared directly. In the original formulation of the paradox it was envisaged that there were twins, one who underwent this return journey and the other who didn't. On the traveler's return one can't have both twins younger than each other, so which one really has now aged more than the other, or are they still both the same age?

The experimental answer is provided by the experiment we mentioned earlier involving the muons traveling repeatedly around the circular path. These muons are playing the part of the astronaut. They start out from a particular point in the laboratory, perform a circuit, and return to the starting point. And it is these traveling muons that age less than an equivalent bunch of muons that remain at a single location in the laboratory. So this demonstrated that it is the astronaut's clock which will be lagging behind the mission controller's when they are directly compared for the second time.

So does this mean that we have violated the principle of relativity and revealed which observer is *really* moving, and consequently which clock is *really* slowed down by that motion? No. And the reason for that is that the principle applies only to inertial observers. The astronaut was in an inertial frame of reference while cruising at steady speed to the distant planet, and again on the return journey while cruising at steady speed. But—and it is a big "but"—in order to reverse the direction of the spacecraft at the turn-around point, the rockets had to be fired, loose objects lying on a table would have rolled off, the astronaut would be pressed into the seat, and so on. In other words, for the duration of the firing of the rockets, the craft was no longer an inertial reference frame; Newton's law of inertia did not apply. Only one observer remained in an inertial frame the whole time and that was the mission controller. Only the mission controller is justified in applying the time dilation formula throughout. So, if he concludes that the astronaut's clock runs slow, then that will be what one finds when the clocks are directly compared. Because of that period of acceleration undergone by the astronaut, the symmetry between the two observers is broken—and the paradox resolved.

At least it is *partially* resolved. The astronaut knows that she has violated the condition of remaining in an inertial frame throughout, and so has to accept that she cannot automatically and blindly use the time dilation formula (in the way that the mission controller is justified in doing). But it still leaves her with a puzzle. During the steady cruise out, she is able, from her calculations, to conclude that the controller's clock was going slower than her own. During the steady cruise home, she can conclude that the controller's clock will be losing even more time compared to her own (the time dilation effect not being dependent on the direction of motion—only on the moving clock's speed relative to the observer). That being so, how on earth (literally) did the mission controller's clock get *ahead* of her own? What was responsible for *that*? Is there any way the astronaut could calculate in advance that the controller's clock would be ahead of hers by the end of the return journey? The answer is yes; there is. But we shall have to reserve the complete resolution of the twin paradox for later—when we have had a chance to see what effect acceleration has on time.

Length Contraction

Imagine the spacecraft traveling to a distant planet. Knowing both the speed of the craft, v, and the distance, s, from the earth to that planet, the mission controller can work out how long the journey should take as recorded on his clock. He finds $t = s/u$. The astronaut can do the same kind of calculation. But we already know that her time, t', will not be the same as the controller's—because of time dilation. So, won't she find that she has arrived too soon—that she couldn't possibly have covered a distance, s, at speed, v, in the reduced time, t'? That would allow her to conclude that it must be she who is really moving. This would again violate the principle of relativity. Something is clearly wrong. But what? It cannot be the speed, v; both observers are agreed as to their relative speed. No, the resolution of the dilemma lies with their respective estimates of the distance from the earth to the planet. Just as the controller has his time, t, and the astronaut

On returning from a trip into space, the astronaut finds that the journey time as recorded by the mission controllers' clocks is longer than it was according to her clock. This photograph shows Mission Control in the Johnson Space Center in Houston, Texas, during the landing of space shuttle *Discovery* on September 11, 2009.

has hers, t', he has his estimate of the distance, s, and she has hers, s'. How do they differ? In the same ratio as the times differed:

$$(2) \qquad \text{For the astronaut} \qquad s' = vt'$$

$$s' = vt\sqrt{(1 - v^2/c^2)}$$

$$\text{But for the controller} \qquad s = vt$$

$$\text{Therefore} \qquad s' = s\sqrt{(1 - v^2/c^2)}$$

In other words, the astronaut is perfectly happy about her arrival time at the planet. The reading on her clock is less than it is on the controller's because, according to her, she has not traveled as far as he claims she has done. At a speed of $0.67c$, the journey time according to her is $^4/_5$ of what he says it is because she holds that she has traveled only $^4/_5$ the distance. Thus her estimates of time and distance are perfectly self-consistent—just as the controller's set of estimates are internally self-consistent.

In this way we come across a second consequence of relativity theory. Not only does speed affect time, it also affects space. As far as the astronaut is concerned, everything that is moving relative to her is squashed up, or contracted. This applies not only to the distance between earth and the planet, but to the shape of the earth itself, and of the planet itself; they are no longer spherical. All distances in the direction of motion are contracted, leaving distances at right angles to that motion unaffected. This phenomenon is known as *length contraction*.

And, of course, from the principle of relativity, what applies to the astronaut, applies also to the controller. Distances moving relative to him will be contracted. At the speed with which the craft is traveling, $0.67c$, the length of the moving craft will appear to the controller to be only $^4/_5$ of what it was when stationary on the launch pad. And not just the craft, but all its contents—including the astronaut's body; she will appear flattened (see figure 5). Not that she will feel it. This is not the sort of flattening one gets when a heavy weight is placed on the chest, for instance. It is not a mechanical effect; it is space itself that is contracted. This kind of contraction affects everything, including the atoms of the astronaut's body; they will be reduced in size in the direction of motion—and hence they do not need as much space to fit into her body. So she feels nothing. Neither does she *see* that everything in her craft is squashed. This is because the retina at the back of her eye is squashed in the same ratio, so the picture of the scene cast onto the retina takes up the same proportion of the available area, and hence the signals to the brain are as normal. All this applies at whatever speed she travels. Right up close to the speed of light, the spacecraft could be flattened thinner than a CD, with the astronaut inside and still not feeling a thing, and seeing nothing unusual.

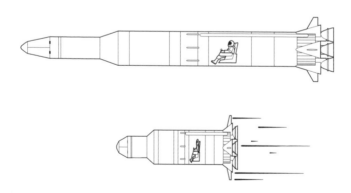

5. According to the mission controller, not only the speeding spacecraft is length-contracted, but all its contents, too.

One final point before leaving the topic of length contraction. Figure 5 illustrates what the controller concludes about the spacecraft as it speeds past him; it is length-contracted. But is that what he actually *sees*—with his eyes? Is that what a photograph of the craft would look like? Here we must take account of the finite time it takes light to travel from the different parts of the craft to the lens—the lens of the controller's eye or of a camera. If the craft is approaching him, for instance, light from the nose cone has less distance to travel than light from the rear and so will take less time. But what we see on the photograph is made up of light that has all arrived at the same time. That being so, the light that makes up the image of the rear of the craft must have been emitted earlier than that which goes to make up the image of the nose cone. So what he sees, and what is on the photograph he takes, is not what the craft was like at a particular instant, but what different parts of the craft looked like at different instances. The picture is distorted. It so happens that the distortion makes it appear that the craft is rotated—rather than contracted. It is only when one takes into account the different journey times for the light making up different parts of the picture that one can calculate (note that word "calculate" again) that the craft is not really rotated but is traveling straight ahead, and that it is length-contracted.

Loss of Simultaneity

We have seen how relative speed brings about time dilation and length contraction. There is a further way in which time is affected. Recall the experiment where a pulse of light was fired at right angles to the direction of motion of the spacecraft and its arrival at a target placed on the ceiling of the craft was timed. Let us imagine another experiment. This time the astronaut takes two sources of pulsed light. Both sources are placed at the midpoint of the craft. One is directed toward the front of the craft, and the other toward the rear. They point at targets placed at equal distances from their respective source. The two sources each emit a pulse at the exact same instant (see figure 6a). When do the pulses arrive at their targets? The answer is obvious. The pulses travel identical distances. They both travel at the normal speed of light, *c*. So they arrive at their destinations simultaneously (see figure 6b). That is the situation as seen from the perspective of the astronaut.

But what does the mission controller conclude when he observes what is going on in the craft as it speeds past him? This is illustrated in figure 7. Like the astronaut, he sees the two pulses leave their sources at the same time—simultaneously (figure 7a). Next he sees the rear-going pulse strike the target at the back of the craft. What about the

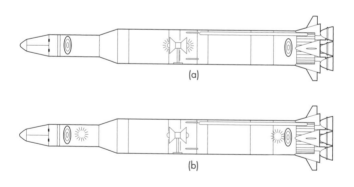

(a)

(b)

6. According to the astronaut, two pulses of light emitted at the same time from the center of the spacecraft will arrive at the ends of the craft simultaneously.

forward-going pulse? According to the controller, this pulse has *not* yet reached its target; it still has some way to go (figure 7b). Why the difference? From his perspective, the rear-going pulse has less distance to travel because the target placed at the back of the craft is moving forward to meet its pulse. In contrast, the forward-going pulse is having to chase after its target which is tending to move away from it. Both pulses are traveling at the same speed, *c*. So, the rear-going pulse will arrive at its destination in a shorter time. The forward-going pulse arrives some time later (figure 7c).

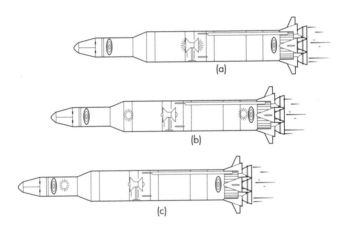

(a)

(b)

(c)

7. According to the mission controller, the two pulses emitted at the same time from the center of the spacecraft do not arrive at the ends of the craft simultaneously.

Thus we find that whereas the two observers are agreed about the simultaneity of events that occur at the same point in space (the two pulses leaving from the midpoint of the craft), they do not agree about the simultaneity of events separated by a distance—the arrival of the pulses at the two ends of the craft. For the astronaut the events were simultaneous; for the controller the rear-going pulse arrived first. Indeed, one might add that from the perspective of a third inertial observer in a spacecraft that was overtaking the first one (and so from that perspective the first craft would appear to be going backward), it would appear that the pulse directed at the front of the craft arrived first—before that directed to the rear—which, of course, is quite the reverse of what the controller on the ground concluded.

That appears to raise a particularly worrying problem—to have two events such that observers disagree as to which one happened first. Suppose, for example, the two events consisted of (i) a boy throwing a stone, and (ii) a window breaking. Might there not be a perspective from which the window breaks before the stone has been thrown?!

Fortunately this paradoxical scenario is not possible. The order of two events that could be causally related is never reversed; all observers perceive the cause to have occurred first regardless of their motion relative to the events. As you have probably heard (and we shall be dealing with this later), nothing can travel faster than the speed of light. For event A to be the cause of event B, it must be possible for a signal, or some other kind of influence, to pass between them at a speed that does not exceed that of light, c. If that is the case, then observers, while disagreeing as to the time interval between the two events, will agree over the order in which the events occurred. Only when one is dealing with two isolated events that can have no influence on each other can there be disagreement over the order in which they occur. So, in summary, where causality is concerned, there is no paradox.

But that still seems to leave us with the question as to who is right? Are events such as the arrival of the two pulses at the targets in the spacecraft *actually* simultaneous or not? It is impossible to say; the question is meaningless. It is as meaningless as asking what the *actual* time of the journey from the earth to the planet was, or what the *actual* length of the craft was. The concepts of time, space, and simultaneity take on meaning only in the context of a specified observer—one whose motion relative to what is being observed has been defined.

Space-Time Diagrams

All this talk about the loss of simultaneity and the question of causality can perhaps be made clearer with the help of a diagram such as that shown in figure 8. It is called a *space-time diagram.* Ideally we would like to be able to draw a four-dimensional representation of the three axes of space and one of time. But that, of course, is impossible on a flat two-dimensional sheet of paper. So we suppress two of the spatial axes by fixing our attention on events occurring along only one of the spatial directions: the x' axis. This might, for example, be a line joining the front and back of the spacecraft along which the light beams passed in that experiment exploring simultaneity. The second axis shown on figure 8 is one representing time. In point of fact, it is customary to label this ct' rather than t' as this enables both directions on the diagram to be measured in the same units—units of distance. All events occurring at time zero will be located somewhere along the x' axis; all events occurring at $x' = 0$ will be found located on the ct' axis.

Let us first consider the loss of simultaneity. The $x' = 0$ coordinate represents the center point of the spacecraft where the two light sources were placed. The two dashed lines represent the trajectories of the two

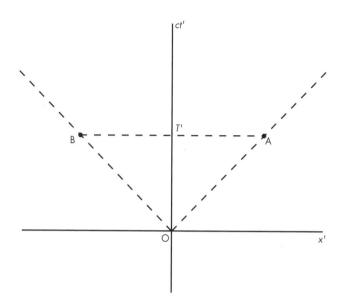

8. This space-time diagram shows the passage of the two light pulses from the center of the craft, O, to the two ends, A and B, according to the astronaut. They both arrive at time *T′*.

light pulses, one going to the front of the craft, the other to the rear. The point O represents the emission of the pulses at $x′ = 0$, $ct′ = 0$. Points A and B mark the arrivals of the two pulses at the two end walls of the craft, having traveled equal distances in opposite directions. A and B are seen to share the same time coordinate, *T′*; in other words, they occur simultaneously. This is the situation as viewed by the astronaut.

How ought we to represent the situation as it appears to the mission controller? In figure 9, the axes labeled *ct* and *x* are those belonging to the controller's coordinate system. All events occurring at the position $x = 0$

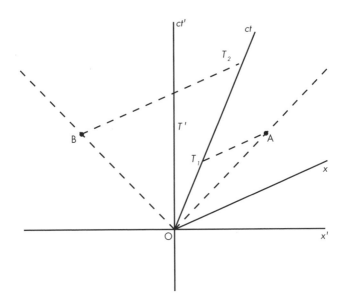

9. This space-time diagram shows how the mission controller's *ct* and *x* axes are inclined to the astronaut's *ct′* and *x′* axes. Although the controller agrees with the astronaut that the two pulses leave the center of the craft simultaneously, at O, according to him, they arrive at the two ends, A and B, at different times, T_1 and T_2.

(for the controller) will occur at progressively different values of *x′* (for the astronaut) because the origin of the controller's coordinate system is moving relative to the spacecraft. Thus, the *ct* axis will be sloping compared to the *ct′* axis. Likewise, the *x* axis slopes compared to the *x′* axis. In other words, the controller's coordinate system is squeezed toward the dashed line of the light pulse trajectory. According to the controller, events occurring at the same time lie along one and the same dotted line running parallel to the x axis. From which we can immediately see how the time coordinate of point A is not the same as that of point B; it is T_1 in one case, and T_2 in the other. The arrival times of the pulses are not simultaneous for the controller—the result we obtained earlier in a somewhat different manner.

What about the question of causality? How is that illuminated by the use of a space-time diagram? As briefly mentioned before, we shall later be showing that nothing can travel faster than light. So, on a space-time diagram, the trajectory of a moving object cannot have a slope flatter than the dashed line representing the trajectory of a light pulse. The line OL in figure 10 represents a possible path of an object such as a ball being rolled along the floor of the spacecraft to the end wall. Likewise, LM is the path of the ball as it returns to the center of the craft having rebounded from the end wall. The line ON, on the other hand, is *not* a possibility for the ball; it would require a speed greater than that of light.

Consequently any event, R, occurring in Region I could have been caused by something happening at point O. This is because it would be physically possible for some influence to pass between the two at a speed which did not exceed that of light. In the case of point L, it was indeed causally connected to O, the influence passing between them being the rolled ball. Likewise, an event at P in Region II could be the cause of what happens at O. All observers are agreed that P lies in the past of O, and that L and R lie in the future of O.

But what of events, such as N, in Region III? There can be no causal link between O and N because, as we have seen, no signal or anything else could travel between the two of them sufficiently fast for one to affect the other. It is events in Region III that are ambivalent as to which one occurs

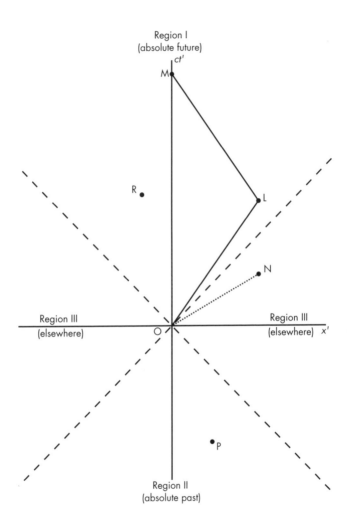

10. This space-time diagram illustrates the three regions in which events may be found—absolute future, absolute past, and elsewhere—relative to the event O.

first. Different observers can arrive at different conclusions depending upon their state of motion relative to the events being observed. But this is of no consequence. The order of causally linked events is never in doubt. All observers are agreed that cause is invariably followed by the effect.

If you are wondering why there are two regions labeled Region III, let me remind you that in this diagram we are depicting only one of the three spatial dimensions. If we wish, we could imagine a second spatial axis coming up out of the plane of the paper. We could then imagine one of the Region IIIs being rotated up out of the plane of the paper, about the *ct′* axis, and being overlaid on top of the other Region III. Thus, the two Region IIIs are one and the same region. Similarly, we could imagine the dashed line of the light pulse trajectory being rotated about the *ct′* axis, tracing out a cone. Indeed, this is referred to as the *light cone.* Region I, contained within the light cone, is said to lie in the *absolute future* of the point O; Region II, also contained within the light cone, is in the *absolute past* of point O. As for Region III, that carries the name *elsewhere*(!).

Another common term used in connection with space-time diagrams is *world line.* Again, it is a rather odd name. It refers to the line traced out on a space-time diagram depicting the path of an object or light pulse. In figure 9, for example, the lines OA and OB are the world lines of the two light pulses traveling from the center of the craft to the front and back. In figure 10, the combined path OLM represents the world line of the rolling ball. As you sit reading this book you are yourself tracing out a world line. If you are at home, you are considered stationary, maintaining the same position coordinates. But time is passing. Your world line will therefore be one that is parallel to your time axis. If you are reading this book on a train, then to someone observing your train passing by, you are changing both your position coordinate and time coordinate. In that observer's reference frame, your world line will be inclined to his time axis much like that of the rolling ball. As the train slows down, it will become more closely parallel to the time axis.

Four-Dimensional Spacetime

All this talk about different observers having different perceptions about space and time can be disorienting. One occasionally hears people claiming that relativity theory can be summarized in the phrase "all things are relative"—implying that it's a free-for-all and anyone can believe anything they want! Nothing could be further from the truth. Observers might not assign the same values for time intervals and spatial distances, but they do agree about how their respective values are related—through the formulae we have derived for time dilation and length contraction. These are determined with mathematical rigor.

Not only that, there is a measurement about which all inertial observers can agree. Let me explain. In ordinary, everyday life we are happy to accept that if someone were to hold up a pencil in a room full of people, everyone would see something different. Some would see a short-looking pencil, others a long one. The appearance of the pencil depends on one's viewpoint—whether one is looking at it end-on or broadside-on. Do these differing perceptions worry us? Do we find them disconcerting? No. This is because we are all familiar with the idea that what we see is merely a two-dimensional projection of the pencil at right angles to our line of sight (see figure 11). What one sees can be captured on a photograph taken by a camera at the same location, and photographs are but two-dimensional representations of objects that actually exist in three spatial dimensions. Change the line of sight and one gets a different projected length, p, of the true length, l, of the pencil. We are happy to live with these different appearances because we are aware that when one takes into account the extension of the pencil in the third dimension—along the line of sight—then all observers in the room arrive at the same value for the actual length of the pencil—the length in three dimensions. Those who are viewing the pencil end-on, and thus see a short projected length, have to add in a large contribution for the component of length along the line of sight; those viewing broadside-on with a long projected length have little to add in the way of the component along their line of sight. Either way, they arrive at the same value for the true length in three dimensions.

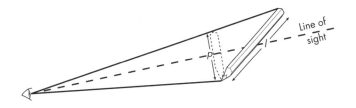

11. A pencil of length *l* has a projected length, *p*, at right angles to the line of sight of an observer.

We use this as an analogy for explaining our differing perceptions of time and space. In 1908, three years after Einstein had published his special theory of relativity, one of his teachers, Hermann Minkowski (who once described his distinguished student as "a lazy dog"), approached the subject from a different angle and suggested a reinterpretation. He proposed that what relativity was telling us is that space and time are much more alike than we might suspect from the very different ways in which we perceive and measure them. Indeed, we should stop thinking of them as a three-dimensional space plus a separate one-dimensional time.

Rather, they were to be seen as a four-dimensional *spacetime* in which space and time are indissolubly welded together. The three-dimensional distance we measure (with a ruler, say) is but a three-dimensional projection of the four-dimensional reality. The one-dimensional time we measure (with a clock) is but a one-dimensional projection of the four-dimensional reality. These ruler and clock measurements are but *appearances*; they are not the real thing.

The appearances will change according to one's viewpoint. Whereas in the case of the pencil being held up, a change of viewpoint meant changing one's position in the room relative to the pencil, in spacetime, a change of viewpoint entails both space and time and consists of a change in speed (which is spatial distance divided by time). Observers in relative motion have different viewpoints and therefore observe different projections of the four-dimensional reality.

Hermann Minkowski (1864–1909), a German mathematician, taught Einstein at the Eidgenössische Technische Hochschule Zürich, also known as the Swiss Federal Institute of Technology. Minkowski's book *Raum und Zeit* (*Space and Time*) was published in 1907.

What is being proposed here is that space-time diagrams, such as figures 8 to 10, are not simply to be regarded as graphs of spatial distances plotted against time intervals. Where graphs are concerned, one is free to plot any variable one chooses against any other. Space-time diagrams do that, but they have an added significance: they represent a two-dimensional slice taken through a four-dimensional reality.

What is the nature of this four-dimensional reality? What are the contents of spacetime? These will depend on the three dimensions of space and the one dimension of time. In other words, they are *events*. Here we must be careful. The word "event" in normal usage can take on a variety of meanings. The Second World War, for example, might be referred to as an important event in world history. "Event" in this context includes everything that constituted the war, spread over the period 1939–45 and wherever it happened. In the present context, however, the word takes on a quite specific, specialized meaning. Events are characterized by their happening at a certain point in three-dimensional space and at a certain instant of time. Four numbers then precisely locate the position of the event in spacetime. One event might be the spacecraft leaving earth at a certain time. A second event might be the arrival of the craft at the distant planet at a different location in space and at a later instant of time. Whereas in three-dimensional space, we are familiar with the idea that the lines join up contiguous spatial points, in spacetime, world lines join up contiguous events.

Our two observers, the astronaut and the mission controller, disagree about "appearances"; i.e., the difference in time between the two events and also the difference in space between the two events. However—and this is the crucial thing—they do agree about the separation between these two events in four-dimensional spacetime, as would all other observers, regardless of their speeds. And it is the fact that all observers are agreed as to what exists in four dimensions that strengthens the idea that spacetime is what is real.

So, what is the distance between events in four-dimensional spacetime? As is well known, in a two-dimensional space, the distance, l,

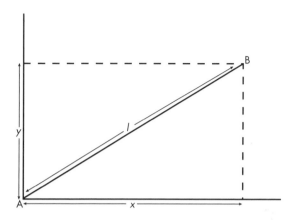

12. A length, _l_, can be expressed in terms of the components _x_ and _y_, according to Pythagoras's theorem.

between two points, A and B, can be written in terms of the projections, _x_ and _y_, along two axes at right angles to each other (figure 12). To do this, we use Pythagoras's theorem once more:

$$l^2 = x^2 + y^2$$
$$l = \sqrt{(x^2 + y^2)}$$

This expression can be extended to cover a distance in three-dimensional space by adding a third term relating to a third axis, _z_, at right angles to the other two:

$$l = \sqrt{(x^2 + y^2 + z^2)}$$

The "distance" or "interval," _S_, between two events in four-dimensional spacetime can be represented through the inclusion of a fourth term related to the fourth axis time, *t*. In order to get the units right (distance being measured in meters and time in seconds), the fourth component has to be written *ct*, so that it too can be measured in meters. A second complication is that, in order for the expression for *S* to be the same for all observers, it has to be defined in such a way that the time and spatial components appear with different signs:

$$(3) \qquad S = \sqrt{(c^2 t^2 - x^2 - y^2 - z^2)}$$

This is the expression that all observers agree upon as the distance between two events in four-dimensional spacetime.

If the first term on the right-hand side of equation 3—the one dependent on time—is dominant, then we say the interval is *time-like*. S^2 is positive and we are talking of a situation where the later of the two events lies in the absolute future of the first event (see figure 10), and thus might be causally connected. If, on the other hand, the spatial terms add up to more than the first term, we say the interval is *space-like*. S^2 is negative, and the later event (if indeed it is the later of the two) lies in the region labeled "elsewhere" in figure 10. Separating the time-like and space-like regions, we have the light cone. On this cone, S^2 for any pair of events is zero.

The idea of reality being four-dimensional is strange and counterintuitive. Even Einstein himself at first had difficulty accepting Minkowski's suggestion—though later he was won over and declared "henceforth we must deal with a four-dimensional existence instead of, hitherto, the evolution of a three-dimensional existence." Not that this is meant to imply that time has been reduced to being merely a fourth spatial dimension. Although it is indeed welded to the other three dimensions to form a four-dimensional continuum, it yet retains a certain distinctiveness. The light cone encircles the time axis, not the others. Absolute future and absolute past are defined in relation to the time axis alone.

Acceptance of a four-dimensional reality is difficult because it is not something that lends itself to easy visualizing—indeed, forming a mental picture of four axes all mutually at right angles to each other is impossible. No, we must dispense with mental pictures and simply allow the mathematics to guide us.

One of the disconcerting features about four-dimensional spacetime is that nothing changes. Changes occur in time. But spacetime is not in time; time is in spacetime (as one of its axes). It appears to be saying that all of time—past, present, and future—exists on an equal footing. In other words, events that we customarily think of as no longer existing because they lie in the past, do exist in spacetime. In the same way, future events which we normally think of as not yet existing, do exist in spacetime. There is nothing in this picture to select out the present instant, labeled "now," as being anything special—separating past from future.

We are presented with a world where it is not only true that all of space exists at each point in time, but also all of time exists at each point in space. In other words, wherever you are seated now reading this book, not only does the present instant exist, but also the moment when you began reading, and the moment when you later decide you have had enough (perhaps because all this is giving you a headache) and you get up and go off to make a cup of tea.

We are dealing with a strangely static existence, one that is sometimes called "the block universe." Now there is probably no idea more controversial in modern physics than the block universe. It is only natural to feel that there is something especially "real" about the present instant, that the future is uncertain, that the past is finished, that time "flows." All these conspire against acceptance of the idea that the past still exists and the future also exists and is merely waiting for us to come across it. Some leading physicists, while accepting that all observers are indeed agreed on the value of the mathematical quantity we are calling "the distance, or interval, between two events in four-dimensional spacetime," nevertheless deny that we must go that extra step and conclude that spacetime is the true nature of physical reality. They maintain that spacetime is merely a mathematical structure; that is all. They are determined to retain the seemingly common-sense idea that the past no longer exists, the future has yet to exist, and that all that exists is the present. I suspect you are inclined to agree with them. But before lending them your support, it is worth considering in more depth what your alternative to the block universe might be.

It is all very well saying that all that exists is what is happening at the present instant, what exactly do you mean by that? Presumably you mean "me reading this book in this particular location." Fair enough. But I imagine you would also include what is happening elsewhere (literally *elsewhere*) at the present instant. For example, there might be a man in New York climbing some stairs. At the present instant he has his foot on the first step. So, you will add him, with his foot on that step, to your list of existent entities. But now suppose there is an astronaut flying overhead directly above you. Because of the loss of simultaneity of separated events, he will disagree with you over what is happening simultaneously in New York while you are reading this book. As far as he is concerned, the man in New York, at the present instant, has his foot on the second step—not the first step. Moreover, a second astronaut flying in a spacecraft traveling in the opposite direction to the first arrives at a third conclusion, namely at the present instant the man in New York hasn't even reached the flight of stairs yet. You see the problem. It is all very well saying that "all that exists is what is happening at the present instant," but nobody can agree as to what is happening at the present instant. What exists in New York? A man with his foot on the first step, or a man with his foot on the second step, or one who has not yet reached the stairs? As far as the block universe idea is concerned, there is no problem: all three alternatives in New York exist. The argument is merely over which of those three events in New York one chooses to label as having the same time coordinate as the one where you are. Relative motion means one simply takes different slices through four-dimensional spacetime as representing the events given the same time coordinate, "now."

But of course, the block universe idea also has its problems. Where does the perceived special nature of the moment "now" come from, and where do we get that dynamical sense of the flow of time? This is a big unsolved mystery, and might remain that way for all time. It does not seem to come out of the physics—certainly not from the block universe idea—but rather from our *conscious perception* of the physical world. For some unknown reason, consciousness seems to act like a searchlight scanning progressively along the time axis, momentarily singling out an instant of physical time as being that special moment we label "now"—before the beam moves on to pick out the next instant to be so labeled.

But now we are venturing into the realms of speculation. Let's get back to relativity . . .

The Ultimate Speed

We have seen that the faster one travels, the more time slows down. Reach the speed of light, and time comes to a halt. This appears to raise the question as to what would happen if one were to accelerate still further until one was traveling faster than the speed of light. What would that do to time? Would one go back in time? One hopes not. Such an eventuality could cause all kinds of confusion. Suppose, for instance, one were to go back and accidentally run over one's grandmother—and this before she had had a chance to give birth to your mother. Without you having a mother, how did you get here in the first place!? Fortunately, this cannot happen. As mentioned earlier, nothing can travel faster than light. How does that come about?

According to Newtonian mechanics, an object of mass, m, and velocity, v, has momentum, p, defined by the expression:

$$p = mv$$

To make the object go faster, one has to exert a force on it. According to Newton's second law of motion, the force, F, equals the rate of change of the object's momentum. With m being a constant, this is the same as saying the force equals m times the rate of change of the velocity, which is the acceleration, a. Thus,

$$F = ma$$

From this equation we can conclude that if one pushes long enough and hard enough, the acceleration will continue indefinitely, and there will be no limit to the velocity that can be reached.

The notion of time travel has inspired writers, artists, and filmmakers to create some of their most enduringly popular works. On the television series *Star Trek*, in a 1967 episode, Captain Kirk and Spock travel back in time. They begin their journey by stepping through a thinking, talking "time portal." When Kirk and Spock return, having successfully accomplished their mission, the portal says, "Time has resumed its shape. All is as it was before." All good fun, but it does not alter the fact that, although relativity theory allows one to slow time down, it does not allow us to travel back in time.

But this is not how it is in relativity. Just as we had to modify our notions of time and length, so relativity theory further requires us to redefine the concept of momentum. Accordingly, the relativistic expression for momentum can be shown to be:

$$(4) \qquad p = mv/\sqrt{(1-v^2/c^2)}$$

It is perhaps not altogether surprising that exactly the same factor as appeared in the expressions for time dilation and length contraction, namely $\sqrt{(1-v^2/c^2)}$, has appeared once again. (The mathematics used in deriving this formula, though quite straightforward, are somewhat too lengthy and tedious for inclusion here.)

So how does this affect Newton's second law? The idea of force being the rate of change of momentum is retained, but with the new expression for momentum. This in turn means that the specific formulation of the law, $F = ma$, no longer applies. Whereas before we dealt solely with the rate of change of v (i.e., a), now we have to take account of the rate of change of $v/\sqrt{(1-v^2/c^2)}$. If v is small, then essentially we have the classical, Newtonian situation. But if v is close to c, then v^2/c^2 approaches 1, the expression under the square root sign approaches zero, and the momentum becomes infinitely large. Thus a constant force, while continuing to increase the object's momentum at a constant rate, is now producing hardly any increase in the object's velocity. The velocity of light becomes the limiting case. Hence nothing can be accelerated to a speed equal to that of light.

This in turn means that one can never catch up with a light beam. If a spacecraft has headlights, then no matter how hard the astronaut tries to catch up with the emitted light, the beam will always be moving ahead of the craft. Indeed, the first germ of an idea about relativity theory came to Einstein when contemplating what it would be like to catch up with a light beam. He had in mind a situation where one accelerated up to a speed where one was cruising alongside a light beam, from which point of view, it would presumably look stationary (in the same way as two vehicles cruising alongside each other on the motorway at the same speed appear stationary relative to each other). But Einstein knew from Maxwell's laws of electromagnetism that light, being a form of electromagnetic radiation, *had* to be seen as traveling at speed, c; it could not appear to be stationary. Traveling at speed, c, is all part and parcel of what light *is*. So, not only does the mission controller see the head light beam emitted from the craft traveling at speed, c, relative to him, but the astronaut also will see the beam traveling away from her at the same speed, c. And this despite the fact that, according to the controller, the speed of the beam relative to the craft—obtained in the usual way by subtracting one from the other—is much less. Thus Einstein concluded that there must be something seriously wrong with the way we customarily handle the addition and subtraction of velocities. Velocity being nothing more than spatial distance divided by time, it immediately follows that if we are mistaken over velocities, then we must also be mistaken over the underlying concepts of space and time. And we have already seen where that realization eventually led: time dilation, length contraction, and loss of simultaneity of separated events.

Does the fact that we cannot accelerate to the speed, c, rule out all possibility of traveling faster than light? Strictly speaking, no. All we are saying is that it is impossible to take the kind of matter we are familiar with and accelerate it to superluminal speeds. But that does not rule out the rather fanciful possibility of there being a second type of matter, created at speeds exceeding that of light, and which can travel only at speeds in the range c to infinity. Such hypothetical particles have been given the name *tachyons*. Some years ago they were the subject of much speculation. It was noted, for example, that observers made of tachyon matter would consider that speeds in the tachyon world were confined to be less than c, and that it was our type of matter that would have speeds lying in the range c to infinity. But enough of that; there is absolutely no evidence for tachyons; they are but the subject of unfounded speculation.

$E = mc^2$

How are we to interpret the relativistic expression for momentum (equation 4)? Some physicists prefer to think that there is nothing to interpret as such; one merely replaces the v in the Newtonian formulation with the more complicated one, $v/\sqrt{(1-v^2/c^2)}$, retaining the concept of an unchanging mass, m. That is probably the currently most favored position adopted by physicists. However, there remains much to commend an earlier alternative way of looking at things. According to this other viewpoint, the new factor,

$$1/\sqrt{(1-v^2/c^2)}$$

ought to be thought of as belonging to the mass. In other words, mass increases with velocity, v, by this ratio. Such an idea requires us to draw a distinction between the mass of the object when at rest (its so-called *rest mass*), and its mass when moving.

Consequently, the m in the formula should be replaced by the symbol m_0 referring to what the mass of the object is when it is at rest, i.e., with $v = 0$. Thus

$$p = m_0 v / \sqrt{(1 - v^2/c^2)}$$

or

$$p = mv$$

where

$$(5) \qquad m = m_0 / \sqrt{(1 - v^2/c^2)}$$

m now being taken to denote the mass of the object at speed v.

To what are we supposed to attribute this mass increase? As the object increases its speed so it also increases its energy; it acquires kinetic energy—energy of motion. Energy is assumed to possess mass. The object cannot take on the extra energy without at the same time taking on the extra mass that goes with that kinetic energy. Why is there a speed limit? Because the mass, m, of the object eventually approaches infinity as v approaches c, and it becomes impossible for a force, no matter what its magnitude and for however long it operates, to significantly accelerate an object of infinite mass.

We have arrived at this conclusion regarding there being a limiting speed on the basis of theoretical reasoning. But is it borne out in practice? To answer this, we go once more to the high-energy physics laboratory at CERN on the outskirts of Geneva, Switzerland, or to any of several such laboratories in the USA and Europe. There we have machines called *particle accelerators* (popularly, though somewhat erroneously, known as "atom smashers"). Their function is to use powerful electric forces to accelerate tiny subatomic particles— protons or electrons—to high speed. In some accelerators the particles are guided by electromagnets around a circular path—somewhat like an Olympic hammer-thrower whirling the hammer repeatedly around his head as it goes faster and faster. And sure enough, it is found that there is a speed limit—the speed of light. As one continues to push on the particles, their speed creeps up ever closer to that of light, but never quite reaches it—this despite the momentum continuing to increase to the point where it is no longer possible for the magnetic field to hold the particles on course. That then becomes the energy limit of that particular machine. To reach higher energies still, one must build a bigger machine—to accommodate additional magnets. The largest to date, sited at CERN, has a circumference of 27 kilometers.

Interpreting this result as being due to the mass of the particles increasing, how heavy do they get? At an accelerator at Stanford, California, they accelerate the lightest subatomic particles, electrons, down a straight tube three kilometers long. By the time the particles emerge at the other end, they have a mass 40,000 times larger than when they began their journey. Having acquired such a mass, what happens to it subsequently? As the electrons are eventually brought to rest, they lose the energy they once possessed, and, in the process, the mass associated with that energy; they revert back to their normal rest mass.

At this point, an interesting question arises. We have seen how energy—energy of motion—is associated with mass. But what about the rest mass, m_0, the particle possesses when it is stationary and has no kinetic energy? If it is the case that one cannot have energy without the mass that goes with the energy, does it not also suggest that one cannot have mass without there also being energy? If so, what kind of energy is associated with rest mass? The answer is that it is a locked-up form of energy. It is the energy which, under certain circumstances, can be partially released and is the source of the power of nuclear bombs and of the sun.

Examining this in more detail, we note that, just as there is a relativistic expression for the momentum of an object, so there is one for the total energy, E, of an object. It is Einstein's most famous equation:

$$(6) \qquad E = mc^2$$

or

$$(7) \qquad E = m_0 c^2 / \sqrt{(1 - v^2/c^2)}$$

This 2004 photograph taken by the U.S. Geological Survey shows the Stanford Linear Accelerator, located near Interstate 280 and Sand Hill Road in Menlo Park, California. The photograph is an orthoimage—an aerial photograph that has been digitally corrected so that there is no geometric distortion.

The expression can be written

$$E = m_0 c^2 (1 - v^2/c^2)^{-\frac{1}{2}}$$

which, as you might know, can in turn be approximated by

$$E \approx m_0 c^2 (1 + \tfrac{1}{2} v^2/c^2 + \ldots)$$

$$E \approx m_0 c^2 + \tfrac{1}{2} m_0 v^2 + \ldots$$

The first term on the right-hand side represents the energy locked up in the rest mass. The remaining terms represent the additional energy acquired through the particle's motion. The first of these you will recognize as the familiar Newtonian expression for kinetic energy, it being a good approximation to the relativistic kinetic energy for values of v small compared to c. So what we are saying is that the total energy of the object is the sum of the energy locked up in the rest mass of the object, plus the kinetic energy.

In effect, the equation $E = mc^2$ is telling us that a mass, m, is always associated with an energy, E, and vice versa, an energy, E, is always associated with an accompanying mass, m. (The c^2 factor is there in order to get the mass and energy units right; one cannot have, say, E kilowatts = m kilograms!) Thus we can assert that a plate that has been warmed in the oven will be heavier than when it was cold. This is because, being warm, it now has more energy, and therefore must have acquired the additional mass that goes with that energy. Not that such a difference would be noticeable. (So, if you drop the plate on removing it from the oven, the reason will be more to do with the need to use oven gloves than its increased weight.)

But when dealing with powerful forces, such as those that bind atomic nuclei together, it is altogether another story. In nuclear processes, mass differences become significant. As you doubtless know already, atoms consist of a heavy central nucleus surrounded by very light electrons. The 92 elements that make up all the matter we find in nature differ from each other in the number of electrons they have (ranging from 1 to 92) and also in the size of their nuclei. It is found that light nuclei in collision with each other sometimes fuse together to form a heavier nucleus. As with all bound systems, once the composite nucleus has formed it would take energy to prise the components apart again. From which we conclude that the two smaller nuclei must have had more energy between them initially than when they were later combined within the larger nucleus. The act of combining must, therefore, have required the energy difference to be released. This is done in the form of heat energy and/or the energy of light. Such then is the process whereby the sun gets its energy—*nuclear fusion*—the fusion of light nuclei to form larger nuclei.

The larger nucleus, possessing less energy than its earlier separated components, must also have less mass than the separated particles. Some of the energy originally locked up in the form of rest mass energy has now been transformed into other manifestations of energy, which subsequently get radiated out into space. In this way, the sun converts 600 million tons of hydrogen into 596 million tons of helium with the loss of four million tons of rest mass every second.

What of *nuclear fission*? This is the process that powered the first nuclear bombs dropped on Hiroshima and Nagasaki, and is the source of energy for today's nuclear power stations. It depends on the fact that very large nuclei, such as uranium, tend to be unstable. Their neutrons and protons can be packed more tightly and efficiently if the big nucleus were to split to form smaller nuclei and other fission products such as neutrons, electrons, and light pulses. A typical process involves the isotope of uranium, ^{235}U, absorbing a neutron to become ^{236}U, which then splits to form ^{92}Kr (krypton) and ^{141}Ba (barium), together with three neutrons and a release of energy—the energy of nuclear fission. The neutrons so released can subsequently go on to get absorbed by other ^{235}U nuclei, which also split. Hence a chain reaction is set up. If the series of reactions occurs rapidly, there is an explosion (the nuclear bomb); on the other hand, if activated in a controlled manner, then one has a steady release of energy that can be harnessed for peaceful purposes (nuclear power stations).

There is more energy to be had from the nuclear fusion of hydrogen than from the fission of heavier nuclei. For this reason, hydrogen bombs

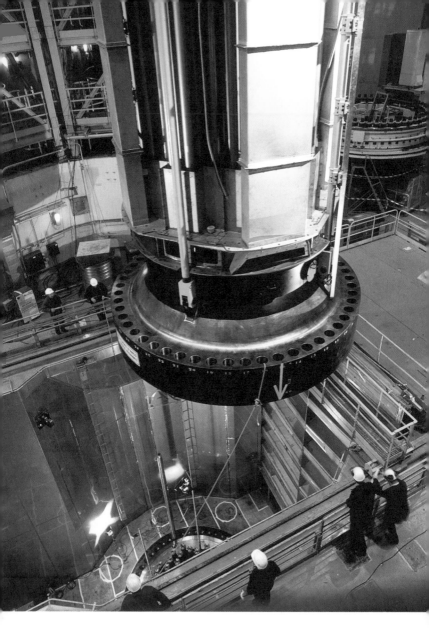

are more devastating than the earlier fission bombs. Ever since the invention of the hydrogen bomb, attempts have been made to harness the power of nuclear fusion for peaceful purposes, one of the attractions being that fuel for such processes would be readily available in the form of the deuterium isotope of hydrogen freely available in sea water. One gallon of sea water contains the equivalent energy of 300 gallons of gasoline. A further advantage of fusion over fission is that it would not result in harmful radioactive waste materials which would then have to be stored safely for enormous lengths of time. Unfortunately, harnessing such power has proved very difficult. The fusing materials have to be at an exceedingly high temperature, 100 million degrees Celsius—so hot it would melt any containing vessel it came into contact with. The material has therefore to be confined by magnetic fields which hold it away from the walls of the container. This is a condition very hard to sustain. Attempts continue, and doubtless one day will prove successful. But the generation of power on a commercial scale still seems a long way off. Current estimates suggest not before the year 2040.

We have seen how rest mass energy can be converted into other forms of energy. Does the process work the other way around? Can, say, kinetic energy be used to create rest mass? Yes, indeed. This is one of the principal aims of the particle accelerators we were talking about just now. The particles are accelerated to high energy and then made to collide with targets, or with a beam of particles traveling in

Generating energy by means of nuclear fission requires a safe method of storing radioactive material. This 2007 photograph shows a containment area inside a containment building—a structure that houses a nuclear reactor, including its fuel rods, and is designed to prevent the escape of radiation. United States government regulations require containment buildings to be able to withstand the impact from a fully loaded passenger airliner.

the opposite direction. One finds that the collisions often produce new particles—particles that were not there initially. The old adage "matter can neither be created nor destroyed" clearly does not hold. Mind you, it is not a case of getting something for nothing. Adding up the kinetic energies of all the final particles and comparing that with the energy originally possessed by the projectile, one finds that some is missing. This shortfall is accounted for by the amount of new rest mass that has been created.

What kinds of particles can be created? In the first place, one notes that one cannot create new matter in any quantities one might like. There are certain fixed allowed masses they can have. Thus one can produce a particle with mass 264 times that of the electron, but not one that is 263 or 265 times the mass of the electron. This is the neutral pion we encountered earlier when discussing the speed of light emitted by a moving source. As we mentioned there, this particle is unstable and decays into two light pulses. Thus, in a short time the kinetic energy of the projectile that was converted into the pion's rest mass reconverts into energy in the form of light. The muon we earlier met in connection with the test of time dilation is another of the new particles arising out of high-energy experiments. It has a mass 207 times that of the electron and results from the decay of a charged pion. The muon in turn decays into lighter particles, once again with the release of energy.

Some of the newly created particles have properties not possessed by the ordinary matter that makes up our world—properties with exotic-sounding names such as *strangeness* and *charm*. This is the realm of high-energy physics, or, as it is sometimes called, fundamental particle physics. It is a world where almost everything is moving at speeds close to that of light, and where special relativity reigns supreme. It is a world where physicists look upon relativity as nothing more than a matter-of-fact, everyday phenomenon—just plain common sense.

That concludes our study of special relativity. Referring back to the preface, you will see how we have already modified five of the so-called common-sense ideas with which we started. What of the others?

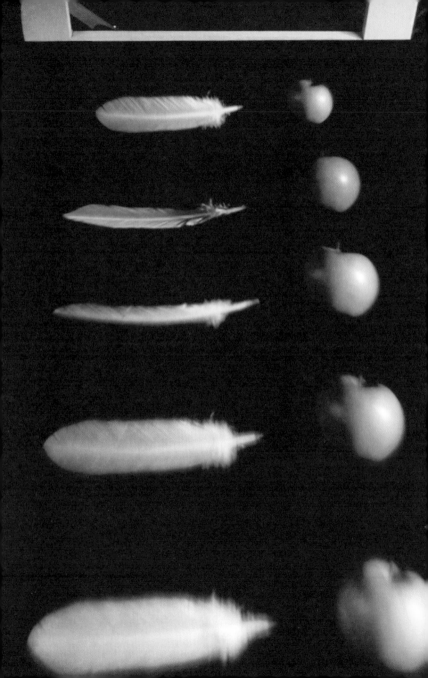

TWO

General Relativity

•

The Equivalence Principle

So far, we have considered only cases where the motion was steady; the observer was in an inertial frame of reference. Also we took no account of gravity. We now move on to broaden our scope to include the effects on time and space of accelerated motion and of gravity. In this wider context we shall see that what we have considered so far, the special theory of relativity, is but a special case of the more general theory.

We begin with the simple observation that in a gravitational field, such as that on the surface of the earth, all objects when released at the same height above the ground accelerate toward the ground at the same rate.

An experiment conducted at Andrews Air Force Base in August of 1988 showed that when an apple and a feather are dropped in a vacuum chamber at the same time, they fall at the same rate. The same type of experiment, using a hammer and a feather, had in 1971 been carried out by the Apollo 15 astronauts on the moon.

Actually this is not immediately obvious. In practice we have to contend with air resistance, which tends to slow down some falling objects more than others. Whereas a hammer falls directly down, a feather released at the same time will float down more leisurely. But when the effects of air resistance are excluded—as was the case when the astronauts on the Apollo 15 mission performed this experiment on the moon—the feather and a hammer arrive at the ground at the same instant.

This is no new insight; Galileo got there before the astronauts. Though the story of him dropping objects from the leaning tower of Pisa is probably apocryphal, he did establish *the universality of free fall*. He did this by carrying out experiments in which objects were rolled down inclined planes. (One should perhaps point out that although skydivers, prior to activating their parachutes, might claim to be in "free fall," they are *not*. They are subject to air resistance.) A statement of the principle of the universality of free fall would go something like this:

> If an object is placed at a given point in space and given an initial veloc-
> ity there, its subsequent motion is independent of its internal structure
> or composition, provided it is subject only to gravitational forces.

So, how are we to understand this? If the acceleration due to gravity is g, then the gravitational force, F, on an object is given by

$$F = m_G g$$

where m_G is a property of the body called its *gravitational mass*.

But, in the Newtonian approximation, we have already seen that the force is also given by the expression

$$F = m_I a$$

where a is the acceleration, and m_I is the *inertial mass* of the object—a measure of the object's inertia when it comes to responding to forces. Eliminating F from these two equations gives us

$$m_G g = m_I a$$

The universality of free fall says that the acceleration, a, of both the hammer and feather are identical. Hence we can talk about *the* acceleration due to gravity, and denote it by g. So, a is identical to g, which means that

$$m_G = m_I$$

and we are able to speak of *the* mass of the object, previously and more usually denoted by m. Experimental tests of the equality of the two types of mass have been carried out to an accuracy of one part in one million million, i.e., 1 in 10^{12}.

As said earlier, this has been a well-known fact for a long time. The genius of Einstein was that, yet again, he spotted that there was something strange going on, which others had overlooked. With special relativity he had noted that there was something odd about trying to reconcile the well-known principle of relativity with the equally well-known fact that the speed of light, derived from Maxwell's laws of electromagnetism, was a constant. Now Einstein found himself puzzled by the fact that these apparently two distinct types of "mass" had the same value. In effect, he was asking how the gravitational attraction "knew" how hard it had to pull on two very dissimilar objects in order to make them accelerate at exactly the same rate. In any case, *why* would gravity want to accelerate them at the same rate? What was the point of that? In this way, he was led to the conclusion that there must be some very close and subtle connection between gravity, on the one hand, and acceleration, on the other.

To see what this connection might be, let us imagine dropping the hammer and feather in an elevator—an elevator being a reference frame that can easily be accelerated in the vertical direction. Suppose at the instant the objects are released, the cable of the elevator is severed so that the elevator itself falls. The elevator would accelerate in exactly the same way as the two dropped objects. They all fall together, meaning that their relative positions do not change. To an observer in the elevator, on releasing the feather and the hammer, they would stay where they were relative to himself. They would not end up on the floor. In other words, it would appear to the observer that gravity had been switched off. The contents of the elevator would be "weightless." (We are assuming that he knows that an emergency brake will eventually come into action, which is why he is able to concentrate on more esoteric physics problems rather than his own safety.)

The idea of weightlessness is more familiarly encountered in the context of astronauts cruising in outer space. It is commonly believed that they are weightless because they are so far out into space that they have gone beyond the pull of gravity exerted by the earth and sun. This is quite wrong. Weightlessness can be experienced while the astronaut's craft is in orbit about the earth. The fact that the craft travels around in an orbit, rather than going off into space in a straight line, immediately tells us that the craft—and the astronaut inside the craft—are being pulled on by the force of gravity exerted by the earth. No, the weightlessness condition arises because the craft is in a state of free fall under the influence of the earth's gravity—just like the observer in the dropping elevator. The reason the craft does not crash down on the surface of the earth is because the earth's gravitational attraction is all being used up simply converting normal straight line motion into the orbital motion we observe; there is none left over, so to speak, to pull the astronaut down on to the earth's surface. Hence the astronaut appears to "float weightless" around the orbit.

Similarly, one can create an artificial "gravity force" by suitably accelerating. Suppose, for example, with the spacecraft cruising and not

Astronaut Stephanie Wilson floats weightless aboard the space shuttle *Discovery* while it is docked with the International Space Station in October of 2007. Floating in front of Wilson is a model of the *Harmony* node, a "utility hub" that astronauts attached to the ISS during the mission. The condition of weightlessness results not from the absence of gravity but rather from the shuttle and the astronaut being in a state of free fall.

requiring attention, the astronaut decides to take a nap. While she is asleep, the rocket motors are activated. On awakening, she feels a pull toward the rear of the craft; any loose objects are seen to be drifting to the rear. What would she conclude? She can hear the rocket motors purring away, so would know that one possibility was that the craft was accelerating. But there is an alternative. What if, when she was asleep, the craft had entered the vicinity of some planet that was now positioned to the rear of the craft, and the rockets were firing merely to maintain the craft's position relative to the planet? If that were the case, the craft would not be accelerating—it would be stationary—and the observed behavior in the cabin would all be due to the planet's gravitational force. It would be impossible for the astronaut to distinguish between the two alternatives: (i) a steady acceleration in outer space, or (ii) being stationary under the gravitational force exerted by a nearby planet. This arises because of the *weak equivalence principle*. This states that one cannot distinguish motion under gravity and acceleration—they are equivalent. As such, the weak equivalence principle is essentially a restatement of the universality of free fall.

Why "weak"? Because there is another version which is called the *strong equivalence principle*. This goes somewhat further and asserts that *all* physical behavior (not just motion) is the same under gravity as for acceleration.

There is one caveat one ought to add. Strictly speaking, one *can* tell the difference between acceleration and gravity. Take a look at figure 13a. The man in the elevator is holding the two objects at arm's length, to the side. The force of gravity is directed toward the center of the earth. Because of their different positions relative to the center of the earth, the force on the hammer is in a slightly different direction from that on the

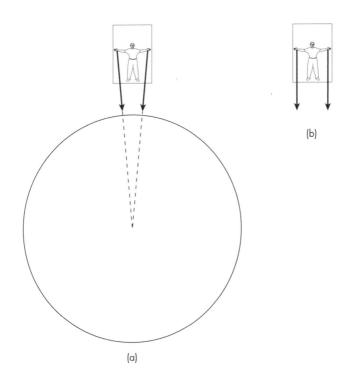

(b)

(a)

13. In case (a), the paths of two objects falling under gravity are slightly inclined to each other as they are both directed toward the center of the earth. In contrast, for case (b), where there is acceleration and no gravity, the paths of the objects are parallel.

feather—the two directions meeting at the earth's center. In contrast, if the observer were to be out in space, far from any gravitating bodies, and accelerates, as in figure 13b, the paths of the two released objects would be parallel to each other; they would not converge to a point. So for the two objects, the acceleration and the gravity force are not quite in the same direction. This means that were the elevator's cable to be severed, the hammer and feather would not remain *exactly* stationary relative to each other and to the elevator, but would move very slightly toward each other, such that were the elevator to plummet through a tunnel to the center of the earth, the hammer and feather would meet. This means that the equivalence principle (both the weak and strong forms) should carry a health warning. The equivalence of acceleration and gravity applies only if one chooses a small enough region and makes one's measurements to only a certain limited accuracy. Over a larger region and/or to a higher precision, one might begin to see the slight deviations we have been talking about. One ought also to specify that the measurements should not be taken over too long a time. Two objects released from slightly different heights within an orbiting (free fall) spacecraft will, after a sufficiently long period of time, drift apart relative to each other because the force of gravity (which falls off as the inverse square of the distance from the center of the earth) will be slightly less for the object placed higher.

However, this is all something of a quibble. The important thing is that because of the equivalence principle, if we wish to investigate what the effects of gravity will be in a given situation, we can, if it's more convenient, think of the gravity being replaced by an acceleration; if, on the other hand, we wish to investigate the effects of acceleration, we can think of it as being replaced by an equivalent gravitational force. The equivalence principle is sometimes regarded as "the midwife" of general relativity—a theory that goes far beyond the principle itself.

The Effects on Time of Acceleration and Gravity

How gravity and acceleration affect time can be explored by once again making use of a source of pulsed light and a target in the spacecraft.

This time, the source is placed at the rear of the spacecraft, and the target at the front (see figure 14). The source is considered to emit a train of pulses at a regular frequency, f. With the rocket engines off, the craft constitutes an inertial reference frame. Under these circumstances, the pulses arrive at the target at the same frequency rate as they were emitted, namely f.

Now suppose that the moment a pulse is emitted, the rocket motors are fired so that the craft accelerates in the forward direction with acceleration, a. If the distance to the target is h, it will take time $t = h/c$ for the light to reach the front of the craft. During this time, the craft will have acquired a speed,

$$v = at = ah/c$$

This is the speed of the target when it receives the pulse compared to that of the source when the pulse was originally emitted. In other words, the target is receiving the light when it is moving away from the source at relative speed v.

Now, as is well known, when dealing with sound waves such as those emitted by the siren on a moving ambulance, or light waves from a moving source, the frequency at reception is different from that at emission.

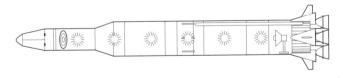

14. This diagram shows a source of light placed at the rear of the spacecraft, emitting regular pulses toward a target placed at the front.

This is the well-known *Doppler shift*. If the source is moving away, then the received frequency is lower; if moving toward, then it is higher. The standard formula connecting the received frequency, f', and the frequency at emission, f, is given by

$$(8) \qquad f' = f/(1 \pm v/c)$$

At speeds close to that of light, this expression should be modified to include the effect of time dilation on the moving source. But for small speeds (such as the speed, v, achieved by the accelerating craft in the time it takes for a pulse to traverse its length) this formulism is sufficient. Rearranging it, the difference in frequency observed at the target as it moves away from the original position of the source can be written

$$(f' - f) \approx - fv/c$$

Using the expression we have derived for v, we finally get

$$(9) \qquad (f' - f) \approx - fah/c^2$$

Thus the frequency with which pulses are received at the front is less than the frequency with which they were emitted at the rear. In similar vein, if the source emitting the pulses were to be placed at the front of the craft and the target at the rear, then the source would appear to be moving toward the observer rather than away, and the frequency with which the pulses are received would be correspondingly higher than at emission.

With this in mind, we now consider what would happen if the acceleration were to be replaced by an equivalent gravitational field.

We suppose the craft to be positioned on its launch pad, held down by the earth's gravity (figure 15). The rear wall of the craft is now to be thought of as the "floor" of the craft, and the front wall as the "ceiling." Again, regular pulses of light pass from the source placed on the floor to the target on the ceiling. Having established what the situation would be for an accelerating frame of reference, we can immediately conclude, from the equivalence principle, that for an observer positioned at the target, the frequency of the pulses arriving there will be judged to be less than what a second observer positioned near the source would judge to be the rate at which pulses are emitted. According to the observer at the top, the frequency with which he receives pulses must equal the rate at which they are emitted. Consequently, he concludes that the frequency of emission is less than that which the observer at the bottom claims it to be. This is called the *gravitational redshift*, as it indicates a shift in frequency to the lower, red end of the spectrum. The significance of this is that if we were to regard the pulsed source as a form of clock—emitting one pulse every second, say—*then the observer at the top concludes that the clock lower down the gravitational field is going slow.*

In similar vein, if the source were placed at the top of the craft and the target at the bottom, then again from the equivalence principle, we must conclude that the observer at the bottom will receive the pulses at an enhanced rate (the equivalent accelerating source coming toward him giving an increased Doppler-shifted frequency). This would be a gravitational *blueshift*. Thus, the observer on the floor agrees with the observer on the ceiling that his clock is going slower than the other.

Note that this is a different kind of conclusion from the one we arrived at over the time dilation phenomenon arising out of relative motion. In that case, both observers believed it was the other person's clock that was

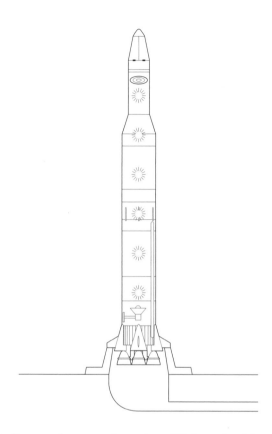

15. This diagram shows regular pulses of light emitted from the rear to the front of the craft, while the craft is standing vertically on its launch pad.

going slow because the situation was exactly symmetrical—there being no way to tell who was "really" moving. This new situation is not symmetrical between the two observers. They are agreed as to which one is really higher up the gravitational field and which one is lower down.

So what we find is that in a gravitational field, a clock—and hence time itself—runs slower the lower down in the field it is. The fractional shift in frequency is the same as we found for the case of the accelerating spacecraft

$$(f' - f)/f \approx -gh/c^2$$

where h is again the difference in height between the two locations, and we have now replaced the acceleration, a, of the spacecraft by g, the equivalent acceleration due to gravity in this uniform field.

Einstein came up with his prediction of a gravitational frequency shift in 1911. The first experimental indications of a gravitational redshift came from a study of the spectra emitted by white dwarf stars. These have a mass of about that of the sun but they are very compact—about 100 times smaller, thus giving rise to a strong gravity field at the surface. More recently, in the 1960s, a team from Princeton were able to measure the shift in the light coming from the sun. But the most dramatic confirmation from astronomical studies involve neutron stars. These have a mass of 1.4 times the solar mass, but radii of only about 10 kilometers. Hence their surface gravity is colossal. In 2002, measurements by the European Space Agency's space telescope XMM-Newton were made of the shift experienced by X-rays emitted by a neutron star and passing through its centimeter-high atmosphere. The shift in frequency was found to be 35%.

In 1960, using an ultra-precise method of measuring frequency, Robert Pound and Glen Rebka experimentally verified the shift by passing gamma radiation up and down a tower of height 22.5 meters. Using the values g = 9.81 m s^{-2}, h = 22.5 m, and c = 3 × 10^8 m s^{-1}, one can verify from the above formula that the fractional frequency shift in this case was only -2.5×10^{-15}. And yet this tiny shift was verified to a precision of 1%.

The effect has also been verified by flying atomic clocks at high altitude in an aircraft. Earlier, we mentioned how the special relativistic time dilation formula was checked using aircraft. In fact, the situation was a good deal more complicated than was indicated there. Two effects come into play: one due to the speed of the aircraft's clock relative to the clock on the ground, and the other—this new effect—due to the aircraft's height above the clock on the ground. These effects are comparable to each other and have to be untangled. In practice, the two experimenters, J. C. Hafele and R. E. Keating, in 1971 flew a clock around the world in an easterly direction, while a second one did the round trip in a westerly direction. The readings on these clocks were compared with a clock at the U.S. Naval Observatory. Though the two aircraft were flying at the same speed relative to the earth's surface, because of the earth's rotational speed they were actually flying at different speeds relative to an inertial observer, say, at the center of the earth. Because of the earth's rotation, the clock on the ground was also moving relative to the inertial observer—with a speed intermediate between those of the two aircraft. For each aircraft journey a log was kept of speed and altitude. This enabled calculations to be made as to the expected loss or gain of the aircraft's clock compared to that on the ground. The clock traveling east should have gained 144±14 nanoseconds due to the gravitational blueshift, but lost 184±14ns due to

The atomic clock system at the U.S. Naval Observatory in Washington, D.C., seen in this 1999 photograph, maintains the official reference standard for precise time on behalf of the U.S. Department of Defense. It is the most accurate atomic clock in the world.

time dilation, yielding a net expected loss of 40±23ns. The experimental result was a loss of 59±10ns. Meanwhile, the westbound clock was expected to gain 179±18ns due to gravity, plus a further gain of 96±10ns due to time dilation, yielding a net gain of 275±21ns. This was also in good agreement with the experimental outcome, which was a gain of 273±7ns.

A further test of the gravitational blueshift was made in 1976 during a rocket flight to an altitude of 10,000 kilometers. Correcting for the expected special relativistic time dilation, the resulting blueshift agreed with theory to two parts in 10^4.

Thus the effects on time due to gravity are well established. Time runs faster upstairs than it does downstairs. But before you get ideas about doing boring jobs, like the ironing, upstairs so that it will be over quicker, do recall that it is time itself that runs faster, not just clocks. This means one's thinking is faster upstairs, so the boring job would still seem to take the same time according to you. It is also worth noting that you will age faster and consequently die quicker up there! Except, of course, the other thing to bear in mind is that the effects we are talking about are negligibly small. Even having climbed to the top of Snowdon, the time it takes to drink a cup of tea in the café there is reduced by only one part in 10^{13} compared to what it would be at sea level.

Not that the gravitational redshift is always small. As we shall be seeing later, the gravity associated with black holes is so powerful as to be able to bring time to a complete standstill.

The Twin Paradox Revisited

Knowing now about the effects of acceleration/gravity on clocks, we revisit the twin paradox.

Earlier we described how the astronaut twin, having traveled to a distant planet, reversed the motion of her craft so as to return to base in order to carry out an unambiguous, side-by-side comparison of the two clocks. She did this by firing the rocket motors, so causing herself to undergo acceleration. In contrast, during the period of the craft's acceleration, the controller felt nothing. This is how the symmetry between the astronaut and mission controller was destroyed. The controller was, therefore, the only one who had obeyed the requirement of remaining throughout in an inertial frame. For that reason, only his calculation was valid, namely that it would be the astronaut twin who would come back younger than himself.

Assuming the distance traveled between the earth and the planet is h, and the speed of the craft is v, then the time, t_c, recorded on the controller's clock for the complete round trip was

$$(10) \qquad t_c = 2h/v$$

The reading on the astronaut's clock, t_a, according to the controller, was time dilated:

$$(11) \qquad t_a = 2h(1 - v^2/c^2)^{\frac{1}{2}}/v$$

The astronaut agrees with this assessment of the reading on her clock—though for a different reason. According to her, the distance between the earth and the planet (as they are seen to go past her) is length-contracted by the factor $(1 - v^2/c^2)^{\frac{1}{2}}$. Thus she is happy about the similarly reduced time on her clock.

The problem lies over the astronaut's assessment as to what the reading on the controller's clock ought to be when she returns. She argues that the earth and controller are moving at speed v relative to her, so the controller's clock will be time dilated. As far as this goes, she is right. During those parts of the journey characterized by steady motion, she, like the controller, is in an inertial reference frame and is fully justified in regarding his clock as running more slowly than her own. (Here we are ignoring any gravitational effects due to the planet she is visiting.)

But what of the period during which the rockets are firing, the craft is slowing down, and she is no longer in an inertial frame? This retardation amounts to an acceleration in the direction toward earth. Having come to a halt, she must accelerate back up to speed, v, this time toward earth—a continuing period of acceleration in the same direction.

We have seen how the effects produced by an acceleration are the same as those that would be produced by an equivalent gravitational field. We can, therefore, replace the craft's acceleration by an imaginary gravitational field, of uniform strength, considered to be stretching all the way from the craft's present position at the planet to where the controller is on earth. Equation 9, namely, $(f' - f) \approx -fgh/c^2$, gives the observed shift in frequency, $(f' - f)$, of light emitted by a source placed at a distance, h, lower down in a gravitational field, g. This is the gravitational redshift. If the source is placed higher up the field, we lose the negative sign in equation 9 and have a blueshift. This relation not only holds for the frequency of emitted light but also for the rate of a clock placed at the same position. Bearing in mind that in our case the controller's clock is placed higher up in the gravitational field compared to the observer (the astronaut), the astronaut concludes the controller's time is speeded up. Thus, for the duration of the acceleration, the astronaut considers that the controller's clock is running faster than her own. This speeding up of time is so pronounced, by the time she switches off the rocket motors prior to cruising home, the controller's clock, instead of lagging behind, is now far ahead of her own. During the steady cruise home, she once again regards the controller's clock to be running slower than hers because of the usual time dilation. As a consequence, during the homeward journey her clock is tending to catch up with controller's. However, it turns out that the latter gained so much time during the short acceleration period,

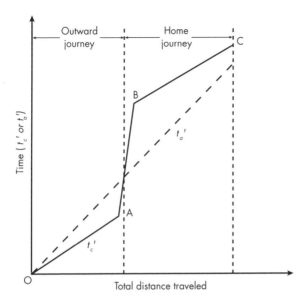

Outward journey | Home journey

Time (t_c' or t_a')

Total distance traveled

16. This diagram shows the reading on the astronaut's clock, t_a', compared to the reading on the controller's clock, t_c', as judged by the astronaut.

it is still ahead of hers when she arrives back at earth. In other words, the stay-at-home twin is now older—which, of course, is the same conclusion as that reached by the other twin. Hence there is no paradox.

The readings on the two clocks, t_c' and t_a', at the various stages of the journey, as judged by the astronaut, are illustrated in the graph of figure 16. Starting out from earth at point O, the craft reaches the planet at point A with t_c' lagging behind t_a'. Between A and B, the rockets are fired, after which t_c' is ahead of t_a'. During the section B to C, the gap between the two readings tends to close. But at C, t_c' is still ahead of t_a'.

One thing that might puzzle you is that, were the craft to undertake a longer journey, say ten times longer, then the time differences would be ten times greater. However, it takes exactly the same acceleration to reverse the speed, v. How can the identical acceleration produce ten times the change in the controller's clock reading? The answer is there in equation 9, where we see that the frequency shift is proportional to the distance, h. Make h ten times greater, and the frequency shift is increased tenfold.

Another concern you might have is that we have not specified how rapidly the acceleration should take place. Again, this is of no consequence. We know that $v = gT$, where g is the acceleration and T is the time the acceleration operates in order to produce that change of velocity, v. Were the acceleration to be halved, it would have to operate for twice the time to produce the same change in speed. Equation 9 shows that with half the value for g, the frequency shift would be halved. But the acceleration, and hence the enhanced running rate of the clock, will continue for twice as long, so the overall change of the clock's reading will be the same as before.

It is easy to check out the matter quantitatively making use of the Doppler shift method. (This is done in the rest of this section, but if you prefer, you can skip it and move straight on to the next.)

Let us assume that the controller's clock emits pulses of light at a frequency of one per second (as reckoned by the controller). The astronaut, by counting the pulses of light received from the other's clock, will be able to keep tabs on the other's clock.

How many pulses will she have received by the time she gets home?

As stated earlier (equation 8), the standard formula connecting the received frequency, f', and the frequency at emission, f, for light emitted by a source traveling at speed v is given by

$$f' = f/(1 \pm v/c)$$

At speeds close to that of light, this expression should be modified to include the effect of time dilation on the moving source:

$$f' = f(1 - v^2/c^2)^{1/2}/(1 \pm v/c)$$

$$f' = f(1 - v/c)^{1/2}(1 + v/c)^{1/2}/(1 \pm v/c)$$

Thus, when the source is moving away from the observer,

$$(12a) \quad f' = f(1 - v/c)^{1/2}/(1 + v/c)^{1/2}$$

And toward the observer,

$$(12b) \quad f' = f(1 + v/c)^{1/2}/(1 - v/c)^{1/2}$$

From equation 11 we see that according to the astronaut, the outward journey takes $t_a/2 = h(1 - v^2/c^2)^{1/2}/v$.

The number of pulses, n_o, received during this outward journey is the time $t_a/2$, multiplied by the frequency of the received pulses (expression 12a)

$$n_o = f' t_a/2 = f(1 - v/c)^{1/2}h(1 - v^2/c^2)^{1/2}/v(1 + v/c)^{1/2}$$

$$n_o = fh(1 - v/c)/v$$

Similarly, the number of pulses, n_r, received during the return journey is the time $t_a/2$, multiplied by the frequency of the received pulses (equation 12b)

$$n_r = f' \, t_a/2 = f(1 + v/c)^{\frac{1}{2}} h(1 - v^2/c^2)^{\frac{1}{2}}/v(1 - v/c)^{\frac{1}{2}}$$

$$n_r = fh(1 + v/c)/v$$

The total number of pulses received, n, is given by

$$n = n_o + n_r = fh(1 - v/c)/v + fh(1 + v/c)/v = 2fh/v$$

Given that the frequency, f, is one pulse per second, we arrive at the total time on the controller's clock as *2h/v*.

This is in agreement with the controller's own estimate, as given in equation 10. In this way, the astronaut can anticipate the extent to which the controller's clock will be ahead of hers.

The Bending of Light

We have already seen, through the equivalence principle, how acceleration and gravity produce equivalent effects on the motion of disparate objects such as hammers and feathers. But what of the motion of a light beam? We are accustomed to thinking of light traveling in straight lines, but is this the case under the influence of gravity, or in an accelerating frame of reference?

To investigate this, imagine yet another experiment involving the pulsed light source and target on board the spacecraft. This time, the source and target are arranged to be exactly the way they were in the first experiment. In other words, the beam of light is to be fired at right angles to the direction of motion of the craft.

While the craft is considered stationary and far from any gravitating source—or equivalently, if it is in free fall—it constitutes an inertial reference frame. Under these circumstances, the beam of light, as expected, travels in a straight line to the target; see figure 17a. But now suppose that, at the moment the pulse leaves the source, the rockets are fired and the craft accelerates forward. As far as the mission controller is concerned, the pulse of light again follows exactly the same path— a straight line in the same direction as before. But by the time it reaches the far wall, the craft will have moved forward; the bull's-eye of the target is no longer directly opposite where the source was when the pulse started its journey. In other words, the controller will see it strike a point some-what to the rear of where the target is now.

Meanwhile, what does the astronaut see? This is illustrated in figure 17b. The pulse starts off in the direction of the target, but then in order for it to strike the far wall to the rear of the target, it must deviate from a straight line, following a curve.

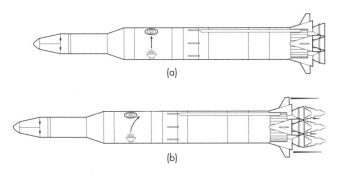

17. For a spacecraft in free fall, case (a), a pulse of light directed laterally across the craft travels in a straight line to the target on the opposite wall. For a craft undergoing acceleration, case (b), the pulse appears to the astronaut to follow a curved path, striking the opposite wall to the rear of the target.

Were we now to think of the acceleration as being replaced by an equivalent gravitational field, where the rear wall of the craft is once again regarded as the "floor" and the nose cone as the "ceiling," the astronaut would conclude that the light pulse had "fallen" toward the floor—much as an object thrown across the craft would start out aiming at the target, but would fall toward the floor and miss the bull's-eye.

From this observation, we would thus expect light rays to follow curved trajectories in gravitational fields; the light would be bent. Indeed, such was the prediction made by Einstein in 1915 while he worked in Berlin during the First World War. News of his ideas got out of Germany to the British scientist Arthur Eddington, based in Cambridge. Six months after the war ended, in May 1919, Eddington verified Einstein's theory through one of the most famous experiments of all time. The idea was to note the normal positions of the stars in a particular region of the night sky. Then the positions were again measured when the sun was in that region. Under the latter conditions, the starlight would have to pass close by the sun to reach us and would therefore have to pass through the gravitational field of the sun. The light would follow a bent path and so, by the time it was detected, it would be coming from a somewhat different direction from its original one. This in turn would give the appearance that the position of the star had shifted from where it was usually to be found; see figure 18. Of course, one problem in making this observation is that the sun's brilliance would normally make it impossible to see the stars. For that reason, the observation was carried out during a total eclipse. The effect being sought was extremely tiny—no more than a deflection of 1.75 seconds of arc (a few ten-thousandths of a degree). But Eddington successfully verified the prediction.

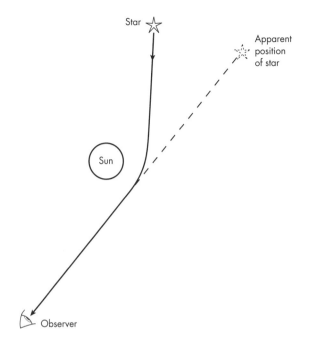

Star

Apparent
position
of star

Sun

Observer

18. The path of light from a distant star changes direction as it passes the sun. By the time it reaches an observer, it appears to be coming from a different part of the sky; the star's apparent position has changed.

This, and later eclipse expeditions, were able to manage measurements of the effect to no better precision than 20%. But then during the period 1989–93, the European Space Agency's satellite Hipparchos was able to carry out high-precision measurements of star positions. Because it was above the earth's atmosphere, the stars were permanently visible, and there was no need to await eclipses. The bending of light was confirmed to a precision of 0.7%. Whereas the earth-bound measurements had had to concentrate on starlight that just grazed the sun's limb, where gravity was at its strongest, and the bending effect was most marked, Hipparchos was able to detect the bending of light even for stars situated at 90° to the sun's direction.

The bending of light gives rise to an interesting phenomenon called *gravitational lensing*. Not just the sun but also a galaxy, or indeed a cluster of galaxies, can act as a gravitational source bending and distorting the light coming to us from a distant luminous object lying beyond it. In 1979, an observation was made of what appeared to be two identical quasars close to each other (quasars being very bright distant sources of light located in massive early-type galaxies). They turned out to be two images of the same quasar. The light from this single source had been distorted by a galaxy which lay along the line of sight to the quasar. The intervening galaxy acted as a kind of lens, bending the quasar's light. If the source, the lensing galaxy, and ourselves were exactly on the same line, then the light from the source would bend uniformly around the galaxy producing a ring—sometimes called the Einstein ring. But that is the ideal situation. Owing to being somewhat off-line, and the lensing galaxy not being spherically symmetric, one more normally sees distorted images and multiple images. This is called strong lensing, and to date over a hundred examples are known. In addition, one can get micro-lensing where a single star acts as the lens for the light from another more distant star lined up with it. In such cases, one sees the light from the source suddenly brighten up for a while as it goes past the line of sight to the intervening star, the latter acting as a magnifying glass. Indeed, in 2004 such a process of magnification revealed that the distant source star had a planet, one-and-a-half times the size of Jupiter, orbiting it. This was the first time an extra-solar planet had been found by this method.

This photograph of the galaxy cluster Abell 370—five billion light years from earth—was taken by the Hubble Space Telescope on July 16, 2009. It is a composite image that uses filters to isolate light from green, red, and infrared wavelengths. The arcs and streaks in the picture are signs of gravitational lensing, a phenomenon that occurs when light is bent by a gravitational field in its path.

It should be noted in passing that Newton, on entirely different grounds to those of Einstein, had much earlier predicted that light would be bent in a gravitational field. He based his ideas on a corpuscular theory of light whereby light was thought of as being made up of a stream of tiny particles traveling at speed, c. Under those circumstances, one would expect the particles to be attracted toward the sun, thus producing a deflection.

However, the amount of the deflection comes out to be only a half of that which Einstein's theory predicts, and which is experimentally verified to be the case. Not only that, but Newton's corpuscular theory was at odds with the wave theory for describing how light moves through space.

Curved Space

So, if Einstein did not regard light as a stream of particles being attracted, much like any other particles, by the force of gravity, what physical picture did he develop to try and understand what was going on here?

We return to the dropping experiment involving the hammer and feather. Given that they have different masses, we saw that the earth's gravity has to pull with different strengths to make them accelerate toward the ground in exactly the same way. This raised the question as to how gravity knew how hard to pull on each to make them behave the same way, and in any case why would it want them to behave like that?

We get the same thing happening when an astronaut goes for a space walk. We consider the spacecraft to be orbiting the earth with its engines off—it is in free fall. The astronaut steps out of the craft and floats there alongside the craft. She too is in orbit about the earth—more or less the same orbit as the craft. Again, the gravity exerted by the earth is such as to produce exactly the same behavior in two very different objects. Instead of traveling in a straight line at constant speed, the force of gravity pulls on the space walker and upon the craft in just the right way to make them both go in a curved trajectory—the same trajectory.

Einstein's response to this was to suggest that in the presence of a gravitating body, the "natural" motion of an object is *not* that of remaining stationary or moving at constant speed in a straight line. Instead, he proposed that near gravitating bodies such as the earth, the space itself becomes distorted. It is curved in such a way that the natural path followed by all objects is whatever path we observe it to be: the orbit followed by the space walker and the craft around the earth.

One way of thinking about this is to imagine a banked race track. On such a track, two very different vehicles can cruise around with little need for the driver to steer because the cars are induced to follow the curved path by the way the track is banked at the corners. The track is distorted or curved in such a manner that it is no longer "natural" for the vehicle to continue in a straight line. It no longer requires a steering force to alter its direction of motion. The "steering" is provided by the shaped track.

So what Einstein is saying is that we do not need to invoke a force—the gravity force—to keep the astronaut and the craft in orbit about the earth. There is no force that needs to be fine-tuned to keep objects of different mass on the same path. Instead, both the astronaut and the craft are just following the natural path that *all* objects will follow if they start out from the same position with the same velocity. Thus Einstein replaced the notion of gravity forces with a completely new conception—that of a *curved space*.

It was simplicity itself. Provided, of course, one can get one's mind around the idea of a curved space! Not easy—especially if one has been brought up to think of space as just another name for "nothing." How can nothing be curved?

The answer is that to a physicist, space is not nothing. Rather, it is to be viewed as a smooth, uniform continuum. Crudely speaking, it is like a very thin jelly (or Jell-O, in Americanese). When we later consider Big Bang cosmology, we shall find that all the galaxy clusters are moving away from each other. This is not because they are spreading out into what was previously unoccupied space—empty nothingness. No, it is more a case of space itself expanding, and in the process, carrying the galaxies along with it on a tide of moving space. Again, when it comes to the study of quantum physics, one discovers that to the physicist space is thought to be jam-packed with "virtual" fundamental particles, some of which pop into fleeting existence from time to time. That is one effect. Another is that the electric charge on an electron, say, repels the charges on the virtual electrons that make up the vacuum close by, so pushing those virtual particles away.

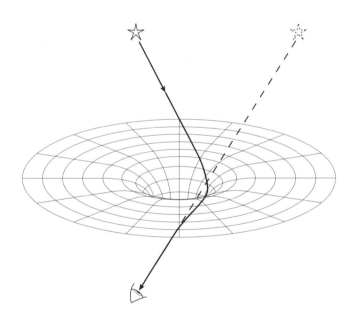

19. This diagram represents the way light from a distant star is bent by the curvature of space caused by the sun.

Thinking in these terms, where space is not nothing but "stuff" (of a sort), it becomes more plausible that the stuff could be distorted and curved in some way so that the natural path to follow would not necessarily be a straight line. And such curvature would be expected to affect everything passing through that region of space—including light. In our previous discussion of the bending of light experiment, for example, we thought of light passing by the sun as being attracted toward it by the force of gravity. This new interpretation, involving curved space, suggests that figure 18 might be replaced by something more like figure 19.

The idea of a curved space is not in itself new. We are all familiar with curved two-dimensional spaces. A two-dimensional space might consist of a flat sheet. In such a plane we find that the circumference of a circle, C, is given by the expression

$$C = 2\pi r$$

where r is the radius. Another result is that the interior angles of a triangle add up to 180°. But we can have a situation where the surface takes the form of a sphere. In other words, the two-dimensional space is curved. Doing geometry on such a surface is very different from what it was on the flat. In figure 20 we see that the circle formed by the equator has as its center the North Pole, P (not the center of the sphere because we are constricted to remaining on the two-dimensional surface). On this surface, the equivalent of a straight line is the shortest distance between two points (the configuration an elastic band would take up if stretched between the two endpoints). Thus, "straight lines" are arcs of great circles for the sphere. Accordingly, PA is a radius, r, of the equatorial circle within the two-dimensional surface (not the radius R from the center of the sphere). The equator is a full circle around the sphere, whereas the radius is but a quarter of a full circle around the sphere. Thus on this surface we have the relationship for this particular circle

$$C = 4r$$

We see that the circumference of the circle is less than $2\pi r$.

Not only circles but also triangles are affected by curved geometry. PAB is a triangle made up of three intersecting "straight lines." The sum

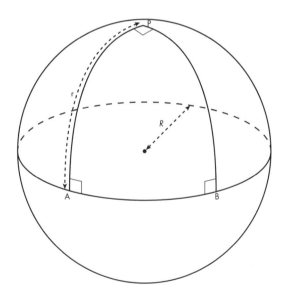

20. Geometry carried out on the surface of a sphere is different from that on a flat surface.

of the interior angles for this triangle is seen to be three right angles; i.e., 270° rather than the usual 180°.

The surface of a sphere is but one kind of curved two-dimensional space. Figure 21 shows another—one based on a saddle shape.

Here we find the interior angles of a triangle add up to less than 180°, and the circumference of a circle comes out to be more than $2\pi r$.

Note that both the circles and triangles, on both types of curved surface, were comparable in size to the overall size of the sphere or saddle. Had we confined our attention to very small circles and triangles we would have obtained quite different results. On the very small scale even a curved surface tends to be pretty flat, in which case the normal geometry for a flat surface holds at least approximately, and that approximation improves the smaller the scale.

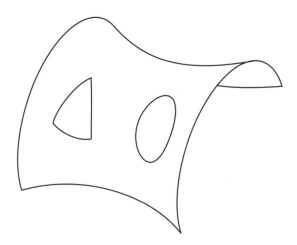

21. The saddle is another shape where geometry carried out within its surface is different from that for a flat surface.

So what we learn from this consideration of curved two-dimensional surfaces is that we get different results from the usual Euclidean flat case, though the smaller in size the figures we use, the closer they approximate to the flat case. These results we carry over into a consideration of what it means to have a curved *three*-dimensional space.

In the first place, it becomes impossible to visualize a curved three-dimensional space. With two dimensions it was easy; we had the third dimension into which we could visualize the curvature going. But where is there a fourth spatial dimension to accommodate the "bend" of the three dimensions?

Actually, visualization can be misleading. Take a look at the surface shown in figure 22. Is it curved? In one sense, obviously yes. It is a cylinder. But looks can be misleading. As far as the *geometry* that goes on in that surface is concerned, it is the same as flat geometry. After all, the cylinder could be made from bending a flat sheet of paper (in a way that you cannot bend a flat sheet to make a sphere or saddle). If you draw a circle or a triangle on a flat sheet, and then bend it to form a cylinder, the properties of those figures remain exactly the same as before.

So forget about visualizing curvature. Instead, we define a space as curved if the geometry carried out *within* that space differs from Euclidean geometry. After all, flies crawling over the surface of the sphere or saddle would not need to have a bird's eye view of the shape of the surface they were on in order to conclude that it was curved. They could arrive at that conclusion simply by performing measurements on triangles and circles within the surface itself. And that is how one explores the geometry of three-dimensional space—not by somehow positioning oneself outside the three-dimensional space for an overall view, but by carrying out measurements within the space itself.

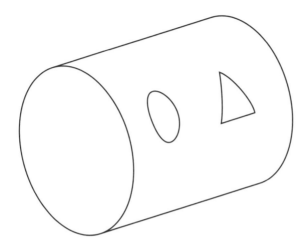

22. Despite the fact that the surface of a cylinder looks "curved," its geometry is the same as that for a flat surface.

From the light-bending experiments, and from the orbiting space-craft and the space walking astronaut, we already know that space gets curved on the scale affected by the earth, sun, and galaxy. These are like scattered dimples on the broad surface that makes up the totality of space. But are they dimples on a surface which, overall, is flat, spherical, saddle-shaped, or what? We shall return to this topic later when considering the universe in general.

Earlier we saw how the presence of a gravitating body affects time (the gravitational redshift). Now we see that it also affects space. Bearing in mind how we were led by special relativity to conclude that space and time constituted a four-dimensional spacetime, we now conclude that we ought not to be thinking solely of a curved space, but rather of *curved spacetime.* The time axis together with the three spatial axes are all affected by the presence of the gravitating body.

We spoke earlier of objects following "natural paths" through curved spacetime. The actual name given to such paths are *geodesics*. A geodesic is the path followed by an object in free fall, that is to say, one that is not subject to any non-gravitational forces, such as electric and magnetic influences (the gravitational effects already being taken into account through the curvature of spacetime). In other words, in general relativity, a geodesic takes the place of the straight line in normal Euclidean geometry or in special relativity. Thus, when the light from a star is bent around the sun, it is following a geodesic.

What is the defining characteristic of a geodesic? In three-dimensional Euclidean space, the analogous straight line is defined as being the path having the shortest distance between two points. In spacetime, a geodesic is defined as that path between the two events characterized by having the maximum *proper time.* Proper time is defined as the time that would be recorded on a clock accompanying the object as it moves between the two points in question. In figure 23, we revisit the twin paradox (briefly this time!). This shows the situation from the point of view of the mission controller. O marks the departure of the astronaut from earth; she travels to the distant planet, arriving at P. She turns around and returns to earth, arriving at Q. The controller meanwhile remains stationary and traces out world line OQ. We have already established that by the time he and the astronaut are reunited, his clock reads more than hers. In other words, his proper time is greater than hers. And this will generally be true. No matter what world line the astronaut traces out between the two points O and Q—for example, the arbitrary path shown passing through S—the reading on her clock will always be less than that on his.

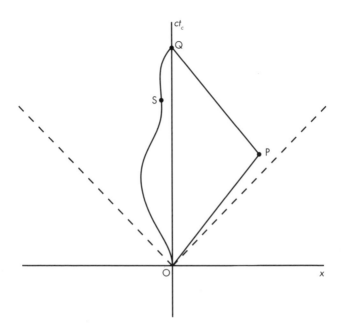

23. This diagram shows world lines for the two twins involved in the so-called "twin paradox."

She will have followed a world line characterized by a proper time that is less than that of the controller. What is so special about the controller's world line that it should have the maximum proper time? Only he remains in an inertial frame of reference throughout; he is following the path of free fall. He is following the geodesic between the two events O and Q.

Incidentally, do not be misled by the rather unfortunate name: *proper* time. It does not mean that somehow this is the actual time, the real time, all other times not being quite right! I reiterate what I said

The American architect R. Buckminster Fuller (1895–1983) is well known for his geodesic structures, one of which is the Biosphere, originally built for the United States pavilion at the 1967 World's Fair in Montreal. The architectural "circles" that go around the circumference at different angles form triangles where they intersect. Unlike triangles on a flat surface, the sum of the angles of these triangles does not equal 180 degrees. Today, the structure is a museum dedicated to water and the environment.

earlier when first introducing relativistic ideas about lengths and times. All estimates of distance and of time are tied to a specific observer's viewpoint. There is no objective distance or time interval independent of any observer's point of view—nothing that can be regarded as *the* distance or time interval.

Another point to note is that, although we have introduced the idea of geodesics in the context of our discussion of the effects of gravity, it applies universally—to cases not involving gravity. It is not a question of using "maximum proper time" in one case and "shortest distance between two points" in the other. In the absence of gravity, the geodesic characterized by maximum proper time happens also to have the property of being the shortest spatial distance.

The crux of general relativity is that matter tells space how to curve, and space tells matter how to move. Space is no longer to be regarded as the passive stage on which the actors—material objects and light—perform their drama. Space itself becomes a performer.

Now you might be thinking that this is all very well: replacing the notion of gravitational forces with that of a curved space or spacetime. But is this not just a matter of personal preference as to how one chooses to see things? Can one not stick with the Newtonian idea of gravity forces if one wishes?

In most everyday situations, Newton's theory holds to a level of precision that is perfectly adequate. Even when computing the orbits of satellites it is fine to use the familiar inverse square law of gravitation. Mathematically, Newton's theory is much easier to handle than general relativity. For this reason alone, physicists will go on talking about gravity forces and will use Newton's law. Nevertheless, they know that the general theory of relativity provides the more accurate predictions and is a superior way of understanding the physics. Newton's law, while being a useful "recipe" for solving most problems—those involving weak gravity and speeds much less than that of light—offers little insight as to what is really going on. General relativity is not just an optional geometric reinterpretation of gravity. We caught a glimpse of this when we pointed out that Newtonian theory did predict a bending of starlight around the sun, on the assumption that light was made up of particles. But it gave the wrong amount. General relativity predicted the right amount.

Another famous test of general relativity was carried out in 1915 and involved Mercury—the planet closest to the sun and therefore able to explore the sun's gravity at its strongest. Like the other planets, Mercury's orbit is an ellipse with the sun at one of the ellipse's foci; see figure 24a. The point of closest approach to the sun is called the *perihelion*. Normally, according to Newtonian mechanics, one would expect the orientation of the orbit to remain unchanged; the perihelion should stay where it is. But it was known that in fact the perihelion of Mercury's orbit tended to change with each successive turn of the planet around the sun (figure 24b). This was called the precession of the perihelion. Most of this movement was easily accounted for in terms of the gravitational attraction of the other planets in the solar system. However, it had been noted since 1845 that the actual rate of the precession differed from that expected by 43 seconds of arc per century. A tiny amount, certainly. But it was definitely there, and was worrying because it was unaccounted for. Except that Einstein's theory did just that. General relativity required just such a precession. Einstein was later to declare that, on hearing the news of the verification of his prediction, he "was beside himself with ecstasy for days."

More recently, in 1974, Joseph Taylor and his research student Russell Hulse discovered pulsar PSR 1913+16 to be a member of a binary system. The pulsar (a form of collapsed star) was in a very eccentric orbit with another star about their mutual center of mass, approaching each other to a distance of 1.1 solar radii at closest approach and retreating to 4.8 solar radii at their furthest separation. And as predicted by relativity theory, it was found that the perihelion is advancing at a rate of 4.2 degrees per year. This is an advance in a single day equivalent to what Mercury does in a century.

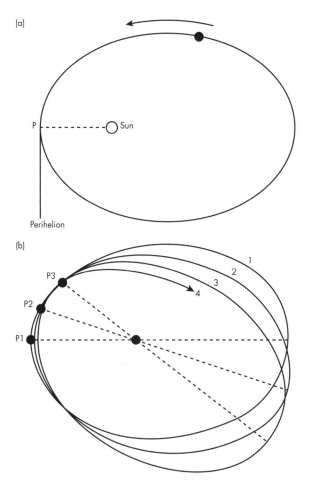

24. According to Newtonian mechanics, a planet such as Mercury should trace out an elliptical orbit. In the absence of any other gravitating body (other planets), the perihelion should remain fixed, case (a). But according to general relativity, the perihelion should precess, case (b).

Another interesting test of general relativity was first proposed in 1964 by Irwin Shapiro and involved using a powerful radar source and bouncing radar pulses off a planet. The idea was to time how long it took for the pulses to travel out to the planet and back, and thus accurately track the planet's path. This was then repeated while the planet was just about to pass behind the sun (see figure 25). Based on the earlier measurements, where the planet was in different parts of the sky, one could calculate what one would expect the reading to be as the radar pulses grazed the limb of the sun. In fact, there turns out to be a time delay of about 250 microseconds. Passing close to the sun causes a slowdown of the pulses. This is what is predicted by Einstein's theory. The experiment has been carried out using pulses bounced off Mercury and Venus, using the planets as passive reflectors. But also using artificial satellites: Mariners 6 and 7, Voyager 2, the Viking Mars Lander, and the Cassini spacecraft to Saturn. In the latter cases, the satellites were used as active retransmitters of the radar pulses. The most accurate experiment to date was in 2003 using Cassini, which was able to verify the prediction to a precision of one part in 10^{-5}.

Note that the effect we are talking about here involves time measurements, and so is a demonstration that it is spacetime, rather than just space, that is curved close to gravitating bodies.

One final point is worth making. We have seen that for all the various tests of general relativity (the gravitational redshift, the bending of light, gravitational lensing, radar probing close to the sun, the precession of the perihelion of Mercury), we were looking for small effects—slight deviations from what would be expected on the basis of Newton's law of gravity. But that should not lull you into thinking that general relativity is concerned only with small, nit-picking matters. General relativity

25. This diagram shows a test of general relativity based on the time delay of radar pulses bounced off a planet as the pulses graze the limb of the sun.

accounts for *all* gravitational effects, including those that can be approximated by Newton's theory. Thus, for example, relativity not only explains the precession of the perihelion of Mercury's orbit, but also why Mercury and all the other planets and satellites are in orbit in the first place.

Black Holes

In figure 19, we tried to illustrate the way the sun curved spacetime by showing it as a ball resting in a hollow it had caused in an elastic sheet.

This, of course, is a very crude analogy. Earlier we noted that when thinking of the curvature of a two-dimensional space, such as the surface of a sphere, then fair enough, one could think of it bending into the third dimension. But when it comes to the curvature of three-dimensional space, there is no additional dimension to take up the "bending." Instead, one must rely on examining the geometric properties of the three-dimensional space itself. Nevertheless, two-dimensional representations of three-dimensional space, like that of figure 19, can occasionally provide some intuitive grasp of what is going on. This is especially so if one has a case of spherical symmetry—like the curvature on the surrounding space produced by the sun—where any two-dimensional slice through that space (which passes through the sun) is representative of any other two-dimensional slice. The third dimension becomes redundant as it contains no information that is not already available from the other two. In the illustration, we can then represent the three-dimensional space by this two-dimensional slice, and use the illustration's third dimension to accommodate the "bending." This is what we did in figure 19. In figure 26, we see how the overall curvature due to the presence of the heavy ball (the sun) causes the smaller ball (a planet) to move around it in orbit rather than move off in a straight line.

In figure 27, we see in more detail the kind of curvature produced by the sun. Why does it have that shape? The steepness of the curve at any point depends on its distance from the center of the sun, and also on how much gravitating matter there is between the chosen point and the center of the sun. As one considers points closer and closer to the sun, the amount of matter remains the same (the mass of the sun), but the distance is reducing, so the steepness of the curve increases. This carries on until one reaches the edge of the sun at point R. Moving into the sun's

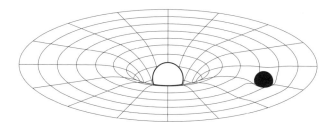

26. This diagram represents the way the curvature of space due to the sun causes the planet to orbit the sun.

interior now, the distance to the center continues to decrease, but now the amount of mass between the chosen point and the center is reducing—an effect tending to reduce the curvature. In fact, the sum of these two effects leads to an overall reduction in curvature such that by the time one has reached the center of the sun, the curve has flattened out. This is what one would expect as the sun exerts no gravity force at its central point. And what is true for the sun, is true of the other stars, and the planets; they create curvatures rather like figure 27.

But again let me emphasize that, although such diagrams might be helpful in visualizing what is going on, in practice we do not see three-dimensional space curving off into some other dimension. Instead, we have to rely on the intrinsic properties of the space itself. So what does that mean? As a specific example we take the case of a spherically symmetric object such as the sun and ask how it affects the spacetime around it.

We already know something about how time is affected. To an observer far from the sun, a clock close to the sun appears to go slow; it is gravitationally redshifted. By how much? Karl Schwarzschild was the first to solve Einstein's equations for the case of a spherically symmetric

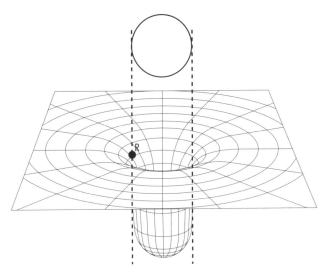

27. This diagram, a profile of the curvature of space caused by the sun, shows how the curvature diminishes within the sun.

body. The solution requires a considerable amount of mathematics, but the end result is fairly simple: for a distant observer, the rate of the clock appears to be reduced by a factor of $(1 - 2mG/rc^2)^{1/2}$. Here m is the mass of the sun, G is the gravitational constant, r is the distance of the clock from the center of the sun, and c is the usual speed of light. We note that for large r, the expression approximates to 1; i.e., when the clock is far from the sun, it appears to go at its normal rate. The closer the clock is to the sun, the slower it goes. For stars heavier than the sun, i.e., with larger mass, m, the effect is greater, as one would expect.

So much for time. How about space? The Schwarzschild solution shows that this is affected in the radial direction. Imagine, for instance, a long line of meter rulers placed end to end extending from the position of our distant observer down toward the sun. According to the observer, the rulers appear to be shortened—the closer a ruler is to the sun, the shorter it is. The factor by which the ruler is contracted is given by the same expression as we had for the slowing down of time: $(1 - 2mG/rc^2)^{\frac{1}{2}}$. Again, we see that for large r the expression approximates to 1 and the ruler appears to have its normal length. The smaller r, or the greater m, the more the ruler is contracted.

What does this do to the speed of light? Imagine a light pulse being emitted from the clock in an outward direction toward the observer. It starts out in a region where time has slowed down. That means everything happening there has slowed down as far as the distant observer is concerned. And that includes the speed of light; it takes longer to cover the distance of each of the meter rulers on its way out from the sun. But not only is time slowed down in the region of the clock, space is squashed up in the radial direction in which the pulse of light is traveling. This means that, according to the distant observer, for each traversal of a meter ruler the light has traveled less than a meter. This is a second factor leading to the slowing down of the light pulse. In effect, the light is having to "drag" itself away from the sun.

A slowing down of the speed of light? But doesn't that violate one of the two postulates upon which relativity theory is founded? No. The postulate specifically spoke about inertial frames of reference, and what we are dealing with here is not an inertial frame. In the curved spacetime produced by gravity, there is nothing to stop the speed of light assuming a value different from the usual c.

So far we have confined our attention to how the situation appears to a distant observer. What of an observer in a state of free fall close to the clock in question? Such an observer is in a local inertial frame of reference. His immediate surroundings appear quite normal. The clock is running at its normal rate, the meter rulers are their normal length, and the speed of light in his vicinity is c. It is important to recognize that just as a small area on the surface of a sphere or saddle approximates to being flat, and the smaller it is, the closer it comes to being flat, so in curved four-dimensional spacetime, if one considers the situation of a freely falling observer in a small local region of that spacetime, then it will appear to be "flat"—meaning that special relativity applies. Thus the curved spacetime around the sun, for instance, can be thought of as being made up of a patchwork quilt of tiny local regions each of which can be dealt with by special relativity. It is only the distant observer who is able to take in the broad, extended picture of what is happening to spacetime both near and far from the sun, and who is able to appreciate its curved features.

In specifying the amount by which the time appears to be slowed down according to the distant observer, and the radial distances are contracted, we spoke of the factor $(1 - 2mG/rc^2)^{1/2}$. It might have occurred to you to wonder what would happen if r were small enough that the second term in parentheses became equal to 1 and the expression reduced to zero. Would that not mean that time would come to a halt and the lengths of the meter rulers would become zero? Here we need to be careful. The Schwarzschild solution (and hence the applicability of that factor) applies only outside where the mass of the sun is concentrated. In other words, beyond point R in figure 27. For the sun, the value of r that would make the factor zero would take one well inside the sun to where only a fraction of the overall mass m would still be contained within the sphere radius, r. So for the sun, the factor can never reduce to zero. However, this is not always the case. There are objects out there in the cosmos that are so compact that the condition can be satisfied. This brings us to the fascinating topic of *black holes*. So, what are black holes, and how are they formed?

We have seen how stars are powered by nuclear fusion processes. But it is clear that, like any other fire, one day it will run out of fuel. What happens then depends very much on how heavy the star is and thus how strong its gravity. For a medium-size star like our sun, after burning steadily for 10,000 million years it will swell up to become a *red giant*. It will shed its outer layers, while the core will collapse down to be a small, bright *white dwarf*. This core will then fizzle out and become a cold cinder.

A star that has a mass of more than about eight solar masses ends its active life with a supernova explosion. Its core collapses down so much under the influence of gravity that the electrons, normally found outside the atomic nucleus, are pushed into the nucleus itself; there they join the neutrons and protons. They then combine with the protons to form more neutrons, and also neutrinos (the released neutrinos being responsible for blasting out material in the explosion). So one is left with a core of neutrons known as a *neutron star*. As mentioned earlier, when dealing with gravitational redshifts, a neutron star typically has a mass of 1.4 solar masses, and yet is only about 10 kilometers in radius. The strength of gravity at the surface of a neutron star is 2×10^{11} that of the earth.

If the initial star starts out with a mass greater than about 20 solar masses, then the supernova explosion results in a neutron star that would have a mass exceeding two solar masses—except that for such a mass, gravity is so strong that nothing can resist it, and the would-be neutron star continues its collapse until all the matter is concentrated

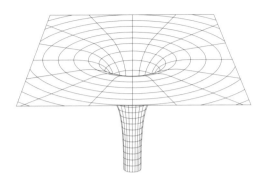

28. This diagram is a profile of the curvature of space caused by a black hole.

down to a point—an infinitesimal region of zero volume and infinite density. This is the birth of a black hole, so named by John Wheeler in the 1960s. The phenomenon was, however, predicted much earlier, in 1939, by J. Robert Oppenheimer and Hartland Snyder on the basis of Einstein's theory.

With regard to the curvature of space caused by a black hole, this is illustrated in figure 28. The curve is to be thought of as continuing down to a singularity at the point where all the matter is concentrated. Thinking in terms of a gravity force, that force would approach infinite strength as we near the center. Anything falling into a black hole gets squashed down to a point at the center. At least, our current physics would lead us to such a conclusion. The worry is that our physics cannot handle singularities. We know that when dealing with very small objects of subatomic size, quantum theory comes into play—and we do not know how to marry up quantum physics with relativity theory. So, nature might have a surprise in store for us. Nevertheless, we have no alternative at the present time but to go along with the conclusion that everything gets squashed down to a point.

That being the case, it follows that unlike the sun, for such an object there will be a distance from the center such that the expression $(1 - 2mG/rc^2)^{\frac{1}{2}}$ reduces to zero. This will be at a radius, k, given by

$$(13) \qquad k = 2mG/c^2$$

k is called the *Schwarzschild radius*. It delineates a spherical surface called the *event horizon*, centered on the concentration of mass. The significance of this distance can be illustrated in the following way. Imagine a spacecraft falling toward the black hole. As it approaches the event horizon, it appears to the distant observer to slow down. This is the combined effect of time slowing down and radial lengths being contracted the closer one is to the center. At the event horizon itself, the craft appears to come to a halt. It appears to be indefinitely suspended there. This is because the light from the craft is having to slowly drag itself away from the region. At the event horizon itself, it takes an infinite time for the light to get back to the observer— hence the object appears to be stationary. Not that in practice it will appear that way for very long. Although the craft appears to the distant observer to have stopped at the event horizon, the craft itself has actually passed through that region quite quickly and carried on into the black hole itself. It emitted only a limited amount of light in its brief transit through that region. So, once that light has crawled out to the observer, there is none left, and the intensity of the light rapidly diminishes and the image fades away.

One has to stress that this is how things appear to be from the point of view of the distant observer. How do things appear from the point of view of the astronaut inside the craft? As far as she is concerned, as she falls toward the black hole, she is initially in a local inertial frame of reference and her immediate surroundings appear normal. There is nothing untoward about time, or distance, or the speed of light. She can pass through the event horizon unaware that from now on her fate is sealed. There is nothing there to indicate that she is passing a point of no return. From now on there is no escape for her. Once inside the event horizon, everything continues in an inexorable plunge toward the center of the black hole. And that is true of light as it is for anything else. Black holes emit no light; hence their name.

The astronaut and her spacecraft end up squashed to a point at the center. And it is important to recognize that this kind of squashing is nothing like the length contraction phenomenon we have come across in the context of special relativity. You will recall that with the length contraction the astronaut in her spacecraft did not feel a thing because all the atoms of her body were contracted and so did not need the same amount of space to fit in. Falling into a black hole, however, would be a totally different matter. Falling feet-first, she would feel her body being stretched lengthwise, as if on a torture rack. This is because her feet, being closer to the center, experience a stronger gravitational field than her head, which is further away. While this stretching is going on, her sides get progressively squashed in. Ultimately she ends up crushed to a point—and very definitely ends up dead!

For a star which ends up as a black hole with a mass, say, ten times that of the sun, equation 13 shows that k would have a value of 10 kilometers. At this distance from the center, the tidal forces acting across a falling astronaut's body at the event horizon would already be colossal. It would be equivalent to being placed on a torture rack where one's feet were attached to a hanging weight of one billion kilograms. Such is the case for a *stellar black hole*—a black hole formed through the collapse of a star.

But that is not the only way black holes form. It is now believed that most galaxies have a black hole at their center—a *galactic black hole*. These are formed through stars near the center of the galaxy being drawn together, colliding, merging, and collapsing down to create a massive black hole. In 1974, our own Milky Way Galaxy was found to have a black hole at its center with a mass of about 3 million solar masses. Most other galaxies appear to contain supermassive dark objects at their centers which are believed to be black holes. Some of these have already swallowed up billions of stars.

From equation 13 we see that the event horizon radius increases in proportion to the mass. The tidal force at the event horizon is known to decrease with the square of the mass. So even for a relatively small galactic black hole containing one million solar masses, the tidal force at the Schwarzschild radius would be reduced by a factor of 10^{12}, meaning that the astronaut would pass through the event horizon hardly affected (though this, of course, is but a temporary respite—the strong tidal forces kicking in at shorter distances).

We have said how stellar black holes are formed when supermassive stars collapse. But one thing we have not yet mentioned is the fact that most stars, like planets, possess angular momentum—they spin about an axis. Angular momentum has to be conserved. So, although some might be lost through the material ejected during the supernova explosion accompanying the collapse of the star, the black hole itself is expected to have to take up much of the original angular momentum.

This complicates matters. The Schwarzschild solution of Einstein's equations no longer holds. It was not until 1963 that Roy Kerr was able to produce the solution for a rotating black hole. The Kerr solution comes up with an especially interesting result: the rotating black hole drags the nearby spacetime itself around like a swirling whirlpool. An object initially falling directly toward the center of the black hole finds itself gradually swept up into this rotating movement. For a rotating black hole, a falling object first passes through a surface known as the *static limit*. This marks the boundary of a region called the *ergosphere* which extends down to the event horizon. The ergosphere is such that the tide of rotating spacetime is so strong that nothing—not even an imaginary spacecraft with an infinite rocket thrust—can remain stationary but must orbit the center of the hole. Only outside the static limit is it conceivably possible for a spacecraft, firing its rockets, to remain stationary.

A space mission named Gravity Probe B is currently trying to test out the prediction of frame-dragging. It consists of four ultra-precise gyroscopes. In free space such gyroscopes would maintain the direction of their axis of spin indefinitely. The probe, however, is in orbit about the earth. The Schwarzschild warping of space caused by the earth's gravity should cause the alignment to change by 0.0018 degrees per year. On top of that, there should be an additional tiny effect due to frame-dragging amounting to no more than 0.000011 degrees per year. That is the equivalent of viewing a human hair from a distance of a quarter of a mile. At the time of writing, we are awaiting results.

When objects fall into a black hole, they lose their identity. For example, being crushed to a point, they no longer have any volume or distinguishing shape. Not that they entirely go out of existence. Whatever mass they had is added to what was already there. What else is retained? The mass of the black hole is one characteristic. Another is angular momentum. Electric charge is conserved, so whatever electric charge was carried by the falling object is retained and added to the total charge on the black hole. And that is it—just mass, angular momentum, and electric charge. All other features of the ingredients that originally went to make up the hole are gone forever.

But, you might be thinking, this is all very well, what is the evidence for the existence of black holes? After all, there is one glaring problem in finding black holes and that is that they are black—they emit no light, and moreover swallow up any light that might otherwise reflect off them. They are to all intents and purposes invisible.

Recall the film of the invisible man. One could not see him directly, but one could see the effects he produced on his surroundings. And that is exactly the approach one adopts when hunting for black holes. One looks for a star that is undergoing periodic changes in the frequencies of the light it emits. This will be due to the Doppler shift as the star first moves away from us and then comes toward us. This motion is characteristic of a binary system consisting of two stars orbiting about their mutual center of mass. Usually one can see both stars. But occasionally there appears to be only the one; its companion is unseen. From the motion of the visible star one can work out the mass of the companion. If this exceeds about three solar masses, then it is a candidate for a black hole. The case is strengthened if the visible star happens to be a red giant; i.e., a star that has a widely distended structure. Sometimes one can see the outer layers of the visible star being drawn across to the invisible companion and emitting X-rays as they are sucked rapidly into the black hole.

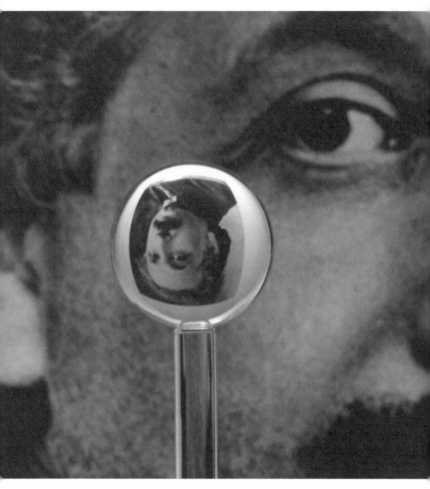

Statistical data from Gravity Probe B, a space mission designed to measure key predictions of Einstein's general theory of relativity, is still being analyzed as of this writing. The mission's satellite, which was launched in 2004 and returned to earth in 2005, carried four gyroscopes made of fused quartz; at the time, these gyroscopes, each about the size of a ping-pong ball, were the most nearly perfect spheres ever created by humans—accurate to within forty atoms of thickness. This photograph of one of the gyroscopes refracts an image of Einstein on its surface.

In 1972, Cygnus X-1 was found by Tom Bolton to exhibit just such behavior. The invisible partner was estimated to have a mass seven times the solar mass. It was a source of X-rays that fluctuated rapidly. The flickering was typically over periods of a hundredth of a second. This period indicated that whatever was emitting the X-rays could not be very big. Light travels only 3,000 kilometers (a quarter of the earth's diameter) in such a time interval, so that appears to set an upper limit on the size of the object emitting the X-rays. In other words, the region is small—consistent with the emission being from the immediate vicinity of a black hole. At the time of writing, there are about 20 known examples of binaries that are best explained in terms of one of the companions being a stellar black hole—some examples being even stronger candidates than Cygnus X-1.

How about the evidence for the supermassive black holes at the center of galaxies? The stars of a galaxy rotate in orbit about the center of the galaxy. Initially one assumed that what was keeping each star on course was the gravitational attraction of all the other stars that were seen to be closer to the center than the orbiting star. However, it was discovered that stars close

This image of the supermassive black hole at our galaxy's center, Sagittarius A*, was made from the longest X-ray exposure of that region to date. During the two-week observation period, Sgr A* flared up in X-ray intensity half a dozen or more times. The cause of these outbursts is not understood, but the rapidity with which they rise and fall indicates that they are occurring near the event horizon, or point of no return, around a black hole weighing in at three million times the mass of the sun.

to the center were orbiting much, much faster than expected on this basis. From this, one concludes that, in order to provide enough attraction to keep the orbiting stars on track, the gravitational mass close to the center must far exceed what could be accounted for in terms of visible stars. This has led to the conclusion that at the very center of the galaxy there must be a supermassive black hole which has swallowed up many stars and hence rendered them invisible.

A second piece of evidence pointing to the existence of supermassive black holes is provided by *active galaxies*. These look like typical galaxies except that they have a small core of emission embedded within them. The output from this core—infrared, radio, ultraviolet, X-rays, and gamma rays—might be highly variable and very bright compared to the rest of the galaxy. This can be explained on the basis of material being accreted by a small central zone—a black hole—with the release of large amounts of gravitational energy.

Added confirmation for the existence of such black holes comes from *quasars*. These are exceedingly bright objects a long distance from us. The further away one looks, the more quasars one sees. As is well known, the further away an astronomical object is, the further back in time we are looking (because of the finite time it takes for light to reach us). Quasars are believed to be galaxies in an early stage of their evolution. As with the active galaxies, the source of the quasars' exceptional brightness was a mystery for some time. But then a connection was made between quasars and the formation of black holes at the center of the newly created galaxies. Indeed, it is now generally believed that, even though active galaxies and quasars look very different to us, they are really the same phenomenon viewed differently. Quasars are simply active galaxies that are very distant from us.

In conclusion, the weight of evidence for the existence of supermassive black holes at the center of galaxies is considered to be overwhelming.

Having dealt with stellar and with galactic black holes, one ought briefly to mention a third possibility: *mini black holes*. We have seen that for an object with mass less than about 2–3 solar masses its gravity is not

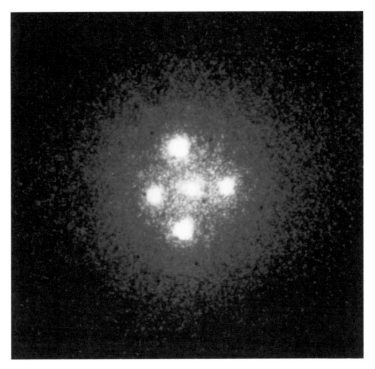

This photograph, taken by the Hubble Space Telescope in September of 1990, shows not five quasars but rather a single one—the image of which is replicated four times (a visual formation known as the Einstein Cross) due to the phenomenon of gravitational lensing (see page 82). In this case, the gravitational body in front of the quasar is the galaxy G2237 + 305, which lies about 122.7 megaparsecs (a megaparsec is about 3.26 million light years) from earth. The quasar beyond it is even farther distant.

strong enough to compress it down to a black hole. However, less massive objects could become black holes if subjected to a sufficiently powerful external pressure. In 1971, Stephen Hawking suggested that under the severe pressure conditions and turbulence of the early Big Bang, perhaps high-density fluctuations did get so compressed as to form mini black holes. These might have had only the mass of, say, a mountain, in which case its event horizon would be no bigger than the size of a subatomic proton. There could be many such objects still around today. However, there is no evidence for their existence.

Likewise, there is no evidence for *white holes*—another theoretical possibility allowed by Einstein's equations. Just as a black hole is a region of space from which nothing can get out, so a white hole would be a region from which one wouldn't be able to stop things spewing out! Another wild speculation, much loved by sci-fi writers, is the *worm hole*. This is the idea that once an object falls into a black hole it gets squirted along a tunnel and out of a white hole somewhere else. This could be somewhere else in this universe, or in another universe altogether. Again, there is no evidence for any such thing.

One last point to note about black holes. Once formed, what happens to them? Do they just stick around forever? For a time they carry on accreting matter and become more massive. But this must come to an end when it has gathered in all the material available to it. One expects that eventually a galactic black hole will have swallowed up all the stars in its galaxy—a process taking of the order 10^{27} years, depending on the galaxy's initial size. Galaxies belong to clusters of galaxies, our own Milky Way Galaxy being one of over 30 members of the Local Cluster. The galaxies are constantly moving about while being bound together by their mutual gravity—somewhat similar to the way a pack of dogs tethered to a stake are free to move about, but within a confined region. As they move the galaxies are constantly emitting energy in the form of gravitational waves (a subject we turn to in the next section). This in turn implies that all the members of a particular cluster will eventually end up together in a black hole. For the Local Cluster, this should take a period of 10^{31} years.

It was originally thought that that was the end of the story. After all, nothing can get out of a black hole, and there is nothing left to go in. But then in 1974, Stephen Hawking came up with the astonishing idea that black holes ought to shine—admittedly very dimly, but nevertheless they ought to emit energy. The reason for this arises out of quantum theory— and therefore, strictly speaking, takes us beyond the remit of this little book. But allow me to sketch briefly how it comes about.

We have already mentioned that to a physicist, empty space—the vacuum—is not empty at all (for one thing, it can get curved). According to quantum theory, the vacuum is constantly, and everywhere, producing pairs of what are called "virtual particles."

These are particle-antiparticle pairs, or pairs of photons (i.e., bundles of light energy). This production of particles requires energy—for example, for the production of the particles' rest masses. But quantum theory allows energy fluctuations to take place; energy can be "borrowed," provided that it is paid back promptly. So, these pairs of particles pop briefly into existence before recombining and going out of existence once more. Hawking suggested that when this process occurs close to the event horizon of a black hole, one of the virtual particles might fall into the black hole releasing gravitational energy (in just the same way as a real particle does when it falls into a black hole). This released energy might be sufficient to pay back the "borrowed" energy without the

Galaxies are gathered together into clusters of galaxies. Our Milky Way, together with Andromeda and about thirty other galaxies, is a member of the Local Cluster. Beyond our cluster, the next nearest one is the Virgo Cluster, pictured here. It consists of about two thousand galaxies.

second virtual particle having to pay back its own energy. This second particle, or photon, just outside the event horizon, is then free to escape the black hole as a normal particle or photon would. Thus, Hawking was led to the conclusion that black holes ought to emit a weak form of radiation. In other words, black holes are not really black. This has come to be known as *Hawking radiation*. It is so weak that it has yet to be observed. A black hole of stellar mass, for instance, would emit radiation equivalent to it having a temperature of only 10^{-7} K above absolute zero.

Nevertheless, most scientists are now convinced that this is how black holes behave. That being the case, it becomes clear that black holes will continually emit energy, and in the process lose mass. In other words, they will evaporate in much the same way as a puddle of water does on a hot day. The smaller the black hole, the more extreme the curvature variation in its vicinity, and the easier it will be for the members of the virtual pair of particles to become separated—one falling into the hole and the other escaping. Hence, the smaller the hole, the more intense the Hawking radiation.

So what is the ultimate fate of a black hole? Black holes of stellar mass are expected to evaporate in 10^{67} years, those of galactic mass in 10^{97} years, and those formed from the amalgamation of all the members of a cluster of galaxies in 10^{106} years.

Gravitational Waves

In the same way as Maxwell's theory represents our understanding of electromagnetism, so Einstein's general theory of relativity is the expression of our understanding of gravity. Maxwell was able to predict, on the basis of his theory, that there should be electromagnetic waves—ripples of electric and magnetic forces spreading out through space. They would be generated by the acceleration of electric charges. Visible light, infrared, ultraviolet, radio waves, X-rays were all examples of such electromagnetic waves; they all travel at the speed of light, differing solely in their wavelength. In the same way, Einstein was able to predict, on the basis of his gravitational theory, that there should be gravitational waves; these would be created by the acceleration of massive bodies. We earlier saw that a massive body such as the sun can be thought of as sitting in an indentation in the fabric of spacetime (see, for example, figure 26). Similarly, gravitational waves can be envisaged as ripples passing through the spacetime fabric. Like electromagnetic waves, they will travel at the speed of light.

Detection of such gravitational waves is no easy matter. This is because the effects they produce are expected to be tiny. In the electromagnetic case, there is no problem. Whirling charged particles around the closed circuit of a particle accelerator (thus subjecting them to centripetal acceleration) readily produces electromagnetic radiation—the so-called *synchrotron radiation*. For electrons, the loss of energy under such circumstances is so pronounced that, in order to reach the highest energies, it is preferable to accelerate them down a tube that is straight, such as the 3-kilometer accelerator at Stanford, California, rather than have them repeatedly guided around a closed circuit.

In the gravitational case, even if one were to whirl a lump of steel weighing several tons at rotational speeds such that it is in danger of flying apart under the centrifugal forces, it would still emit only something like 10^{-30} watts of energy in the form of gravitational waves.

For this reason, we have to look beyond the laboratory, to astronomical objects, for stronger sources of gravitational radiation. The first, somewhat indirect, evidence for gravitational radiation came in 1978. Recall how Hulse and Taylor, four years earlier, had discovered a pulsar that was a member of a binary system. We saw how it was to provide the best test yet for the precession of the perihelion of an orbiting body. There was now to be a second payoff—one that earned its discoverers the Nobel Prize in 1993. Pulsars are neutron stars that emit jets of radiation from their magnetic north and south poles. These jets are then whirled around as the body spins. If we on earth happen to lie in a direction scanned by this rotating beam, then we get a series of regular pulses—much as a ship at sea receives pulses of light from the rotating beam put out by a lighthouse. The beam in this case is of radio waves. What Hulse and Taylor found was that the basic period of this pulsar (0.05903 seconds) was extremely stable (increasing by no more than 5% per one million years), and so provided in effect a very precise clock. Nevertheless, super-imposed on this regular beat was a cyclical variation. This was interpreted as a Doppler shift arising from the way the pulsar moved toward us and then away from us while orbiting its unseen companion. The orbiting period was found to be about eight hours. What was really interesting, however, was that this orbiting period was progressively getting shorter. Not by much—only 75 millionths of a second per year—but over the four-year observational time, the effect was shown to be definitely there. In other words, the pulsar, as it orbited its companion, was losing energy, and was following an ever tighter spiral. This was recognized to be due to it radiating gravitational waves. The calculated rate from Einstein's theory was in agreement with observation to within one half of a percentage point.

But of course, what we would like to do is detect gravitational waves directly by equipment located in the laboratory. Such equipment is shown schematically in figure 29. The idea is to split a laser beam so as to send out two beams in directions at right angles to each other. Having traveled down evacuated tubes for several kilometers, they are reflected back to the origin, where they are allowed to combine and interfere with each other. The idea is that a gravitational wave passing through the detector would cause one of those distances to be increased and the other decreased. This should lead to a disturbance to the way the beams combine—an effect that can be observed with a photo detector. Such an apparatus is called an interferometer. In order to increase the equipment's sensitivity to tiny changes of distance, each beam is made to traverse its return journey

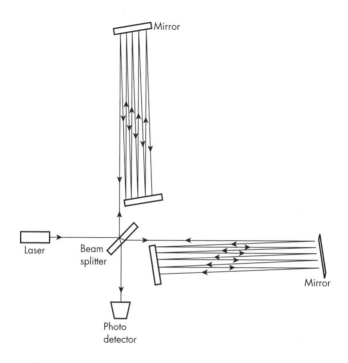

29. This schematic drawing shows the layout of equipment intended to detect gravitational waves.

about 100 times. By this technique it is hoped that one might be able to detect, for instance, the gravitational waves emitted during a supernova explosion. Not that a perfectly symmetrical supernova explosion would be expected to emit such waves. Fortunately, however, they are not expected to be perfectly symmetric. Stars which end their lives in one of these explosions are expected to be spinning. Furthermore, some of them will be members of binary systems. Thus, in practice, supernova explosions are expected to be asymmetric, and will thus emit a pulse of gravitational waves.

The trouble with waiting for supernova explosions is that they do not happen often. In our Milky Way Galaxy they are expected to

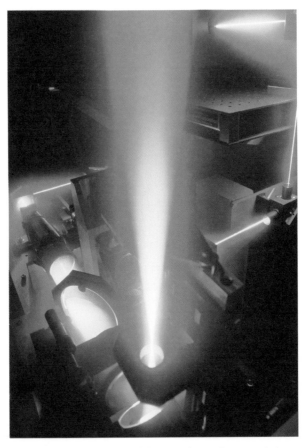

This interferometer emits green argon laser light. It is used primarily to test optical components, including telescope mirrors.

occur on average about once in 30 years. That means there is a good chance of an astronomer spending his whole career waiting for one and ending up with nothing. For this reason, the search has to be extended to other nearby galaxies. But this in turn means, of course, the strength of the signal one is hoping to detect will diminish (the intensity of the signal falling off as the inverse square of the distance). It is the need to be able to detect small signals from other galaxies that sets the degree of sensitivity required of one's equipment. The aim is to detect changes of length of about 1 part in 10^{21} or the equivalent of one-thousandth the size of a proton. Currently there are several of these large interferometers; they are run by U.S., French-Italian, German-British, and Japanese teams. We still await the first positive observation of gravitational waves.

And what of the future? There are already plans to launch an interferometer into space. This would not only have the advantage of being free of the random disturbances here on earth that set limits to sensitivity, it will be able to increase greatly the return path lengths that the laser beams have to travel. The Laser Interferometer Space Antenna (LISA) project envisages three mirror-carrying space stations positioned some five million kilometers from each other. Whereas the equipment currently in operation can, with return paths measured in terms of several kilometers, detect gravity wave frequencies of roughly 100 hertz and more, the extended arms of LISA would allow the detection of much higher frequencies, say, one millihertz. That would allow it to probe the gravity wave spectrum expected to be of interest in investigating the very earliest stages of the universe's evolution—what we shall in the next section be calling "inflation." At the time of writing, the project awaits funding. The earliest launch date would be 2017.

The Universe

Beginning in 1917, Einstein and others began applying general relativity to the universe as a whole. We have already seen from figure 26 how a massive object, such as the sun, causes a local dimple-like curvature of spacetime. It is this that governs the sun's gravitational influence on the motion of the planets. But so far we have given no consideration to the possibility of spacetime having a general, large-scale curvature. As an analogy, consider a mattress. People lying on it will each cause their own indentation. But what if the mattress has an overall sag in the middle? Those in the bed will tend to end up together in the middle. On the other hand, if the mattress has been overstuffed, and people have been in the habit of sitting on the edge of the bed wearing down the springs there, the overall curvature might tend to make the occupants roll away from each other. Of course, the third choice is that one has invested in an expensive orthopedic mattress which remains essentially flat (supposedly very good for us, but rock-hard and uncomfortable). One expects spacetime to behave in one of these three ways. Not only will massive objects cause local dimples, but the *average* mass and energy density will cause an overall, general curvature of spacetime. The equations that deal with this are exceedingly complicated, to the extent that I will not be showing them. Suffice it to say that they only become manageable in the special case where the distribution of matter is both isotropic (the same in all directions) and homogeneous (the same density everywhere). Even so, we shall content ourselves with just a descriptive account of this case. The assumption that our universe is isotropic and homogeneous goes under the name *the cosmological principle*. But is that how the universe is? At first sight it certainly does not appear so. The Solar System is clearly not homogeneous, nor the

Milky Way Galaxy to which it belongs. Nor is the cluster of 30 or so galaxies forming the Local Cluster. There are many other clusters, some consisting of several thousand galaxies. Although the stars within a galaxy, and the galaxies within stars, move relative to each other, they are gravitationally bound—they stay together. Even clusters of galaxies are loosely associated in superclusters. These can take the shape of extended filaments or two-dimensional curved surfaces enclosing voids which contain very little in the way of galaxies. These voids can be as much as 200 million light years across. So even up to this scale, the universe is far from homogeneous.

Fortunately, such distances still represent only a fraction of the size of the observable universe (13.7 billion light years). Thus, one feels justified in accepting the cosmological principle. This being the case, we are faced with three possible alternatives for the overall curvature of three-dimensional space:

(i) It might *be flat,* meaning that, far from any gravitating bodies, ordinary Euclidean geometry would apply. The sum of the angles of a triangle would add up to 180°, and the circumference of a circle, C, would be equal to 2π x the radius, r. Such a space would presumably be infinite in extent.

(ii) Alternatively, it might have what's called *positive curvature.* The two-dimensional analogy of this would be a sphere (see figure 20). The sum of the angles of a triangle would exceed 180°, and for a circle, $C < 2\pi r$. In this case (like the sphere) the universe would be finite in size and closed. This means if one took off in a rocket in a given direction—say, vertically straight up from the North Pole—then after traveling a finite distance, maintaining the same course, one would find oneself back where one started—arriving back at earth at the South Pole. This would be analogous to a fly crawling over the surface of a sphere in a given direction and finding that it had ended up back where it started.

(iii) The third possibility is that three-dimensional space might exhibit *negative curvature*. The two-dimensional analogy in this case would be the saddle (see figure 21). The angles of a triangle would less than 180°, and for the circle, $C > 2\pi r$.

In considering these various possibilities, one might be tempted to think that the correct one is obvious: we *know* that the angles of a triangle equal 180°, and for the circle, $C = 2\pi r$, so space is flat. However, we must remember that even with the analogies of the sphere and the saddle, if we deal only with very small circles, those curved surfaces approximate to being flat. In considering the curvature of the universe as a whole, the only triangles and circles we deal with are tiny and so would be expected to be close to the flat case. In talking of deviations from Euclidean geometry we have to think of triangles, say, that are those involving three very distant galaxy clusters. Only on that kind of scale might we expect to see noticeable departures from flatness.

Which of the three types of curvature applies to the universe depends on its contents. But before coming to that, there is a further observation we must take into account. We have already noted that, according to the cosmological principle, the density of matter everywhere is assumed to be the same throughout space. However, the density does not remain the same over time. As was first observed by Georges Lemaître, in 1927, the universe is expanding. The galaxy clusters are retreating from us. The further away a cluster, the faster it is moving. A cluster that is twice as far away as another is moving twice as fast. This is summarized by *Hubble's law*, proposed by Edwin Hubble in 1929:

$$(14) \qquad v = H_0 r$$

where v is the velocity of recession of the cluster, r is its distance from us, and H_0 is the *Hubble parameter* with a measured value of about 2×10^{-18} sec^{-1}.

This recessional motion is deduced from the way that the spectral wavelengths of the light emitted by distant clusters is shifted toward the red end of the spectrum—the so-called *redshift*. In other words, the wavelengths are stretched out. Initially this was interpreted as a Doppler shift, in much the same way as the sound waves given out by the siren of a speeding police car are shifted to lower frequencies when the car is moving away from us. However, the modern interpretation of the redshift is that it arises from the expansion of space itself. As mentioned briefly before, it is not a case of the cluster moving away from us *through* space. Instead, we envisage the space between us and it as progressively expanding, and in so doing, carrying the clusters away from us on a tide of expanding space. The light does not start out its journey toward us with a wavelength increased by the cluster's motion; rather, it starts out with its normal wavelength, but subsequently this is progressively stretched by the expansion of the space through which it is traveling.

It is important to note that when we talk of space expanding, we do not mean that *all* distances expand. If they did, we would have no

The American astronomer Edwin P. Hubble (1889–1953) inspects the forty-eight-inch Oschin Schmidt telescope at the Palomar Observatory in San Diego County, California, in preparation for the first National Geographic Society-Palomar Observatory Sky Survey, which was begun in 1948 and completed in 1958. The results of the survey are now available as digital images.

way of verifying such an expansion. The binding forces holding together such objects as atoms, the Solar System, galaxies, and clusters of galaxies are sufficiently strong as to overcome the underlying tendency for space to stretch; they thus remain the same size. Not so the weak attraction between the clusters. Here the space-stretching effect is dominant and progressively moves the clusters apart.

This type of recession, where the velocity of recession is proportional to distance, is exactly what one would expect if at some time in the past all the contents were contracted to a point. There was an explosion which blew them apart. This is called the *Big Bang*. The recessional motion we see today is in the aftermath of that explosion. From the observed separation of the clusters at the present time, and the speeds with which they are traveling, we can calculate how much time would be needed for them to have traveled that distance at that speed. That is how one arrives at the conclusion that the Big Bang occurred 13.7 billion years ago.

Hubble's law applies well over moderate distances. Deviations, however, are expected on the largest scale. There is the possibility that the expansion rate will vary over time. Indeed, it was originally anticipated that, because of the mutual gravity operating between the clusters, they would be slowing down. If the average density was sufficiently great, this mutual attraction ought eventually to slow the clusters to a halt. From then onward, they would be brought together once more in a *Big Crunch*. There would be a finite duration between Big Bang and Big Crunch. Not only that, but such a high density would lead to space having a positive curvature, and it would be unbounded but of finite size (like the surface of the sphere in the two-dimensional case).

Of course, with the universe expanding and the distances between clusters increasing, it would be expected that the mutual gravity between them would be reducing. If the density of matter is low, such that the mutual gravity has essentially dropped to zero with the clusters still moving apart, then the expansion would go on forever. In this case we would have negative curvature and a universe infinite in extent.

In between these two extremes lies the so-called *critical density* case. This is where the gravitational attraction essentially drops to zero as the clusters asymptotically approach zero recessional velocity. For this case the geometry is flat. At the current stage of development of the universe, the critical density would have a value of about 10^{-26} kg m^{-3}—equivalent to about 10 hydrogen atoms per cubic meter.

Attempts to measure the rate of deceleration involve observations of the furthest galaxy clusters. There is no difficulty measuring the extent of the redshift. However, there are considerable problems in gaining reliable estimates for the distance to the cluster. For this reason, observational measurements were unable for a long time to gauge the extent of the deceleration and hence distinguish between the three possible models. Then in 1998 came the first surprising indication that the distant clusters were not decelerating at all; they were speeding up! This completely unexpected result revealed the existence of a hitherto unknown type of force—one that acted in the opposite direction to the mutual gravity between clusters, and moreover, at long distances dominated. We shall have more to say about the source of this force later.

As mentioned earlier, the overall curvature of space depends upon the contents of the universe. It was the Russian physicist Alexander Friedmann in 1922, and independently the Belgian physicist and priest Georges Lemaître in 1927, who, using Einstein's theory, developed the equations linking the curvature of space to its source. There are essentially two sources of the curvature of the universe. The first is the average

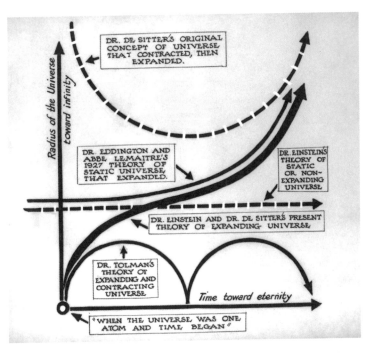

This elegant undated chart positions the then-revolutionary idea advanced in 1927 by the Belgian priest and astrophysicist Georges Lemaître (1894–1966)—that the universe began in an explosion of a single primeval atom—within the context of previous hypotheses.

mass or energy density of the contents of the universe. Here we recall the insight offered by the special theory of relativity that mass and energy are equivalent through the equation $E = mc^2$. An object can be considered to have energy in the locked-up form of rest mass, plus kinetic energy by virtue of its motion. But it is not just matter that has energy. Electromagnetic radiation has energy, as do gravitational fields. So in this context we have to take note of the different kinds of energy there might be. So *energy density* is the first term in the expression for the source of spatial curvature. The second is referred to as *pressure* and arises out of the way clusters are moving away from each other. This concerted motion gives rise to an outward momentum flux which, like energy density, has a contribution to make to the curvature of space. The more important term is the energy density, and that is the one we shall now concentrate on.

So, what do we find? Is the energy density equal to, greater than, or less than the critical value? Adding up the contributions of the visible stars contained in the galaxies, we find it equivalent to about 4% of the critical value. This in itself would indicate that the curvature of space is negative, space is infinite in extent, and the expansion will go on forever. However, we must not be hasty. The sun, like the other stars of the Milky Way Galaxy, are in orbit about the center of the galaxy, held on course by the gravitational attraction exerted by all the matter that is closer in to the center than ourselves. The trouble is that, when we estimate the total mass of those stars, including those swallowed up in the black hole at the center of the galaxy, we find there is not sufficient to exert a gravitational pull strong enough to keep us on our orbit. The conclusion has to be that there is much more matter in the galaxy than that which can be accounted for by the stars. We call this unseen component *dark matter*. What it consists of is uncertain at present, though it is believed not to be the type of matter with which we are familiar—electrons, neutrons, and protons.

Next we note that the galaxies are gravitationally bound together in clusters. Although such galaxies do not orbit each other in the regular way that stars orbit the center of their galaxy, nevertheless the speeds with which they move about within the cluster, without escaping the pull of the other members of the cluster, allows us to estimate the overall mass of the cluster. This comes out to be more than the sum total of the masses of the galaxies themselves, even allowing for the dark matter contained in those galaxies. This in turn implies that there is additional dark matter *between* the galaxies. All in all, it is estimated that the total of all the energy bound up in matter, both visible and dark, amounts to about 30% of the critical density.

Finally, in compiling this inventory of contributions to the overall energy density of the universe, we must take note of the recent discovery that the expansion of the universe is accelerating and why that should be so. Such an acceleration is attributed to an energy density characteristic of the vacuum. At first it seems odd to attribute anything to "empty space." But we have already noted that to a physicist, empty space is not to be thought of as *nothing*. We have already seen how it can be curved, how it can carry galaxy clusters along on a tide of expanding space, and how pairs of virtual particles can fleetingly pop into existence out of the vacuum. This is a possibility allowed by Heisenberg's uncertainty principle. One of its consequences is that at any point in time it is impossible to specify precisely what the energy is. In particular, we cannot specify that the energy of the vacuum is zero—*precisely* zero. This allows the virtual particles to borrow energy on a temporary basis, thus providing them with the energy to produce their rest mass, and hence come into existence. Thus the vacuum is regarded as a seething population of particles coming into existence for short periods of time before disappearing again. This phenomenon gives rise to an average fluctuating energy density for the vacuum—what we now call *dark energy*. It adds its own contribution to the total energy density of the universe. Like the other types of energy, it further increases the overall curvature of space. Where it differs from the other sorts of energy is the way it affects the motion of the galaxy clusters. Whereas the other types of energy give rise to a gravitational attraction, this one gives rise to a repulsion—the repulsion that is responsible for the acceleration of the expansion of the universe.

It is worth noting in passing that in 1917 Einstein himself for a time entertained a related idea. He, like everyone else at the time, was under the impression that the universe was essentially static (the Hubble expansion had yet to be discovered). He therefore needed, in effect, a repulsive force to counter the gravitational tendency to pull all the matter of the universe together. This led him to include in his equation an extra term, called the *cosmological constant,* denoted by Λ. This he was later to regret because otherwise he could have predicted that the

The above image, generated by NASA, shows the development over time of the portion of the universe we can see today since the instant of the Big Bang 13.7 billion years ago. A brief initial period of exceedingly fast expansion, called inflation, was followed by a gradual slowing down due to the mutual gravity operating between galaxy clusters. But now the expansion rate is speeding up due to dark energy. The structure at the right is a representation of NASA's Wilkinson Microwave Anisotropy Probe, launched in 2001, which has provided us with much of our information about the cosmos.

universe was expanding (this being the only other way of keeping the galaxies from coming together).

Although the existence of dark energy has only recently been recognized, it is destined to play the dominant role in the future of the universe. The dark energy density, being a characteristic of the vacuum, remains constant throughout the expansion of the universe. Other forms of energy density, such as those due to matter and to radiation, decrease with the expansion. At first it was the energy density associated with the latter which dominated, and the expansion slowed down. But now those contributions to the overall energy density have dropped below that due to dark energy. As a result, the initial slowing down of the expansion has now been replaced by the observed acceleration due to the dark energy (see figure 30 where we plot R, a measure of the scale of the universe, against time, t). This acceleration is expected to continue into the future.

So, how do we summarize all this? Our current best estimates for the various contributions to the energy density, in terms of fractions of the critical density, are as follows:

Ordinary matter in the form of stars	0.04 ± 0.004
Dark matter	0.27 ± 0.04
Dark energy	0.73 ± 0.04
TOTAL DENSITY	1.02 ± 0.02

That the final result comes out to be so close to the critical value requires explanation. This is because we have to realize that if in the immediate aftermath of the Big Bang the density had been slightly different from the critical value, that difference would by now have become enormously multiplied. If, for example, it had been slightly less than

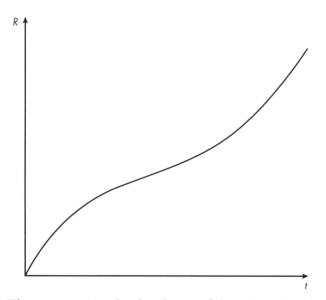

30. The parameter, *R*, related to the size of the universe, is plotted against time elapsed since the Big Bang. At first, the rate of increase of *R* slows down because of the gravitational attraction between the galaxy clusters. But at later times, the contribution due to dark energy dominates, and the rate of increase of *R* accelerates.

critical at the start, then the expansion over the next short time interval would have been greater than that appropriate to the critical density. That in turn would mean a larger volume to be occupied by the energy than would have been so for the critical case. This reduces still further a density that was already too low. Thus the shortfall in density escalates. For example, it has been estimated that if the density today had been found to be 30% of critical, then that could be traced back to a shortfall of only one part in 10^{60} at 10^{-43} seconds after the instant of the Big Bang.

In view of these considerations, it was recognized, even before the recent discovery of the contribution due to dark energy, that the density today was remarkably close to critical. In 1981, Alan Guth came up with a likely explanation of this. He introduced the idea that shortly after the Big Bang there was a period of exceptionally rapid expansion called *inflation*. The universe increased in size by a factor of 10^{30} in a period of 10^{-32} seconds. Whatever the curvature might have been before inflation, afterward it would have been rendered flat. The situation was similar to a balloon being blown up. Although it might have been wrinkled to begin with, afterward, if the expansion has been great enough, any small area of the surface will now be essentially flat. In the same way, our observable universe—that part of the entire universe that lies within 13.7 billion light years of us and thus from which we have been able to receive light emitted since the Big Bang—is but a tiny part of the overall universe. The observable universe is therefore effectively flat.

The conclusion is that, from among the various possible geometries general relativity allows, our universe has a flat space; Euclidean geometry holds. However, spacetime is *not* flat. Because space is expanding with time, the time component is to be thought of as "curved." In this it differs from the spacetime of special relativity, where not only space but also spacetime is considered flat.

In conclusion, we have seen how Einstein's special theory of relativity allows us to understand the behavior of nature's smallest subatomic constituents as they fly about at speeds close to that of light, and how his general theory of relativity provides the essential language and tools for understanding the universe as a whole. Taken together, a remarkable achievement indeed.

FURTHER READING

•

For readers interested in the historical development of the subject:

Jean Eisenstaedt, *The Curious History of Relativity* (Princeton University Press, 2006)

Abraham Pais, *Subtle Is the Lord* (Oxford University Press, 1982)

Books at a similar level to the present one:

Albert Einstein, *Relativity* (reprinted in the Routledge Classics series, 2001 [1954])

Max Born, *Einstein's Theory of Relativity* (Dover, 1962)

Hermann Bondi, *Relativity and Common Sense* (Dover, 1964)

Domenico Giulini, *Special Relativity: A First Encounter* (Oxford University Press, 2005)

Stephen Hawking, *A Briefer History of Time* (Bantam, 2005)

N. David Mermin, *It's About Time* (Princeton University Press, 2003)

Bernard Schutz, *Gravity from the Ground Up* (Cambridge University Press, 2003)

John Taylor, *Black Holes* (Souvenir Press, 1998)

Somewhat more mathematical treatments of the subject:

George F. R. Ellis and Ruth M. Williams, *Flat and Curved Space-Times* (Oxford University Press, 2000)

W. S. C. Williams, *Introducing Special Relativity* (Taylor and Francis, 2002)

Wolfgang Rindler, *Relativity* (Oxford University Press, 2006)

Vesselin Petkov, *Relativity and the Nature of Spacetime* (Springer, 2004)

A full understanding of general relativity requires a sophisticated grasp of mathematics. There are many books written at this level, including the following:

Richard A. Mould, *Basic Relativity* (Springer, 1994)

Robert M. Wald, *General Relativity* (University of Chicago Press, 1984)

Hans C. Ohanian and Remo Ruffini, *Gravitation and Spacetime* (Norton, 1994)

Ta-Pei Cheng, *Relativity, Gravitation, and Cosmology* (Oxford University Press, 2005)

James B. Hartle, *Gravity: An Introduction to Einstein's General Relativity* (Addison Wesley, 2005)

And at the opposite end of the spectrum, for children of ten years and up:

Russell Stannard, *The Time and Space of Uncle Albert* (Faber and Faber, 1989)

Russell Stannard, *Black Holes and Uncle Albert* (Faber and Faber, 1991)

INDEX

•

Note: Page numbers in *italics* include illustrations and photographs/captions.

Abell galaxy cluster, *83*
absolute future/past, *30*, 31, 38
acceleration
    affecting time, 17, *64–65*, 67, 72
    bending light and, *79*
    of expanding universe, 129, 135, 137
    frequency shift and, 76
    of galaxies, 132
    gravity and, *57–63*, 64–72, 74
    speed of light and, 45
    twin paradox and, 74
    unlimited, 41
air resistance, 58
Andromeda galaxy, *119*
angular momentum, 109, 111
atomic clocks, *70–71*

Big Bang, 85, 117, 131, *136*, 137, *138*,
    139
binary systems, 111, 121, 123
black hole(s)
    angular momentum and, 111
    birth of, 106
    causing curved space, *106*

electric charge in, 111
emitting energy, 118, 120
ending of, 117–18, 120
ergosphere in, 110
event horizon in, 107, 108, 109, 110,
    114, 117
evidence of, 111–16
galaxies and, 109, 113–15, *114*
Hawking radiation in, 119, 120
light in, 107, 108
mass and, 111
mini, 116–17
photons in, 118–19
quasars and, 115
rotating, 110
singularities in, 106
spacetime and, 110
from stars, 109, 113
static limit in, 110
tidal force in, 108, 109
time in, 107
block universe, 38–40
Bolton, Tom, 113
bombs, nuclear, 48, 51, 53

Cassini spacecraft, 98
causality, 25, 29–31
CERN laboratory, 4, 11–*13*, 47
charm of particles, 54
cosmological principle, 126, 127, 128
Cygnus X-1, 113

dark energy, 135, *136*, 137, *138*, 139
dark matter, 134–35, 137
Discovery, 61
distance
    contraction of, 20, 103, 104
    curved space and, 100
    four-dimensional spacetime and, 34
    frequency shift and, 76
    simultaneity and, *24*–25
    in spacetime, 35–37
    time and, 17–20, 44
Doppler shift, 66, 67, 76, 111, 122, 129

Eddington, Arthur, 80
Einstein, Albert, *ii*–iv
    bending light and, 80, 84
    cosmological constant of, 136
    curved space and, 84–85
    $E = mc^2$ and, 45–55, 133
    four dimensions and, 37
    gravity and, 59, 69
    Minkowski teaching, 33, *34*
    predicting gravitational waves, 120
    predicting perihelion precession, 96
    relativity principle and, 2
    ring, 82
    speed/velocity of light and, 44
Einstein Cross, *116*
electromagnetism, *3*, 44, 59, 120, 121
elsewhere, *30*, 31, 39
$E = mc^2$, 45–55, 133

energy
    black holes emitting, 118, 120
    dark, 135, *136*, 137, *138*, 139
    density of, 133, 137
    $E = mc^2$ and, 45–55, 133
    kinetic, 46, 48, 50, 53–54, 133
    light and, 51
    mass and, 46, 48, 50, 51, 53
    of motion, 46, 48
    nuclear, 48–49, 50–53, *52*
    rest mass and, 47, 50, 51, 118, 133
    of sun, 48
equivalence principle, 62, 64, 67, 78
ergosphere, 110
European Organization for Nuclear
    Research, *12–13*
European Space Agency, 69, 81
event horizon, 107, 108, 109, 110, *114*, 117

fission, nuclear, 51, *52*, 53
frame-dragging, 110
free fall, 58–59, 60–61, 62, 64, *79*, 84,
    103–4
Friedmann, Alexander, 132
Fuller, R. Buckminster, *94*
fundamental particle physics, 54
fusion, nuclear, 51, 53, 105

galaxies
    acceleration of, 132
    active, 115
    Andromeda, *119*
    black holes and, 109, 113–15, *114*
    clusters, *83*, 117–18, *119*, 128, 129,
        131, 134
    dark matter in, 134–35
    deceleration of, 132
    gravity of, 82–*83*, 131–32, 134, *136*

Milky Way, 2, 109, 117, *119*, 124
    movement of, 85–86
    pressure and, 134
    quasars and, 115
    recessional motion of, 131
Galileo, 2, 58
gamma radiation, 69
geodesics, 92, *94*, 95
geometry, 86, *87*, *88*, *89*, 90–*91*
gravity
    acceleration and, 57–*63*, 64–72, 74
    affecting space, 91
    artificial, 61
    bending light and, 80–82, *81*
    blueshift of, 67, 70, 72, 74
    equivalence principle and, 62
    falling objects experiment on, *56*–58, 84
    free fall and, 60–*61*, 62
    frequency shift and, 65–70, 76
    geodesics and, 92, *94*–95
    inverse square law of, 95
    lensing of, 82–*83*, *116*
    mass and, 58, 84–85
    of neutron stars, 105
    radiation, 121–22
    redshift of, 67, 69, 72, 74, 91, 101, 105, 129
    time and, 64–*65*, 69, 72, 91, 101–2
    waves, 118, 120–25, *123*
    weak, 95
    weightlessness and, 60–*61*
Gravity Probe B, 110, *112–13*
Guth, Alan, 139
gyroscopes, 110, *112–13*

Hafele, J. C., 70
Hawking radiation, 119, 120
Hawking, Stephen, 117, 118, 119

Heisenberg's uncertainty principle, 135
Hipparchos satellite, 81–82
Hubble, Edwin, 129, *130*
Hubble parameter, 129
Hubble's law, 129, 136
Hubble Space Telescope, *83*, *116*
Hulse, Russell, 96, 121–22

inertia, *x*–2, 5, 6, 14, 16, 93, 103, 108
interferometer, 122, 124, *124*

Johnson Space Center, *18–19*

Keating, R. E., 70
Kerr, Roy, 110

Lemaître, Georges, 129, 132, *133*
length contraction, 17–22, *21*, 32, 43, 44, 107, 108
light. *see also* speed of light
    bending, 78–84, *79*, *81*, *86*
    in black holes, 107, 108
    cone, 31, 37, 38
    energy, 51
    time and, *65*, *68*
line of sight, 32, *33*, 82
LISA (Laser Interferometer Space Antenna), 124
Local Cluster of galaxies, 117, 118, *119*, 127

Mariners 6/7, 98
mass
    black holes and, 111
    concentration of, 104, 106, 107
    energy and, 46, 48, 50, 51, 53
    gravity and, 58, 84–85
    increase of, 45, 46, 47, 50

inertial, 59
momentum and, 41
of stars, 105
unchanging, 45
velocity and, 41, 45, 46
matter
affecting space, 95
creation of, 54
dark, 134–35, 137
density of, 128–29, 132
distribution of, 126
Doppler shift and, 76
elements of, 50
tachyon, 45
Maxwell, James Clerk, *3*, 44, 59, 120
Mercury, 96–*97*, 98, 99
Michelson, Albert, 5
microlensing, 82
Milky Way galaxy, 2, 109, 117, *119*, 124
Minkowski, Hermann, 33–*34*, 37
momentum, 41–43, 109, 111
Morley, Edward, 5
motion
energy of, 46, 48
inertia and, 2
law of, 41, 42
of molecules, *3*
recessional, 129, 131, 132
relativity and, *x*–2, 8, 40
space affecting, 95
speed of light and, 4–5, 6
speed/velocity of, 14–15, 41, 45, 46
steady, *x*–2
time dilation and, 67–68
twin paradox and, 73
muons, 11, *12–13*, 15–16, 54
NASA, *136*

neutral pions, 4, 54
neutrinos, 105
neutron stars, 69, 105, 121
Newton, Isaac
bending light and, 83–84, 95
force and, 58
gravity and, 98–99
kinetic energy and, 50
law of inertia and, 2, 16
mechanics of, 96, *98*
second law of motion, 41, 43
Nobel Prize, 121
nuclear energy, 48, 50–53, *52*

Oppenheimer, J. Robert, 106
Oschin Schmidt telescope, *130*

Palomar Observatory, *130*
particle(s)
accelerators, *12–13*, 47, 53–54, 121
charm of, 54
creation of, 54, 135
fundamental particle physics and, 54
muons as, 11
neutral pions as, 4
in nuclear energy, 50–51
properties of, 54
in Saturn's rings, *3*
strangeness of, 54
tachyons as, 45
virtual, 86, 118, 135
perihelion, 96–*97*, 99, 121
photons, 118–19
Pound, Robert, 69
proper time, 92, 93
pulsars, 96, 121–22
Pythagoras, 7, 9, *36*
quantum physics, 86, 106

quasars, 82, 115, *116*
radar, 98, *99*
Raum und Zeit (Minkowski), *34*
Rebka, Glen, 69
red giant stars, 105, 111
repulsion motion, 136
rest mass
    creation of, 53
    definition of, 46
    energy and, 47, 50, 51, 118, 133
    of neutral pions, 54
    nuclear energy and, 48–49
    of virtual particles, 135

Sagittarius A, *114*
Saturn, *3*, 98
Schwarzschild, Karl, 101, 102, 104, 110
Schwarzschild radius, 107, 109
Shapiro, Irwin, 98
simultaneity
    distance and, 22–27, *23*, *24*, *28*
    loss of, 39, 44
    in spacetime diagrams, 26, *27*, *28*
singularities, 106
Snyder, Hartland, 106
space. *see also* space, curved; universe
    affecting motion, 95
    content of, 85–87
    contraction of, 21
    creating virtual particles, 118
    gravity affecting, 91
    matter affecting, 95
    speed/velocity and, *20–21*
    three-dimensional, 90, 92, 100, 101
    time dilation in, *18–19*
    two-dimensional, 87–90, 100
Space and Time (Minkowski), *34*
space, curved, 84–*99*. *see also* space

black holes causing, *106*
dark energy and, 135
geometry for, *88*, *89*, *91*
light bending in, *86*
orbiting planets and, 100–*101*
perihelion precession and, *97*
sun and, *102*
twin paradox and, *93*
space-like, 37
spacetime
    black holes and, 110
    as changeless, 38
    curved, 91–92, 98, 104, 126, 139
    distance in, 35–37
    distinction of, 37–38
    events in, 35
    four-dimensional, 32–40, 91, 104
    geodesics in, 92
    gravity waves and, 121
    light cone encircling, 38
    "now" moment in, 40
    proper time in, 92
    time in, 38
space-time diagrams, 26–32, *27*, *28*, *30*
speed of light. *see also* light
    calculating, 3–4
    CERN laboratory and, 47
    Einstein and, 44
    electromagnetism and, 120
    $E = mc^2$ and, 45–55, 133
    exceeding, 25, 29, 41, 45
    length contraction and, 21
    motion and, 4–*5*, *6*
    slowing, 103
    time and, *8–11*, 22, 41
speed/velocity
    affecting time, *8–11*, 14–15, 17
    definition of, 44

momentum and, 41–43
of motion, 14–15, 41, 45, 46
space and, 20–*21*
Stanford Linear Accelerator, 47, *48–49*, 121
stars, 105, 109, 111, 113
Star Trek, 42–43
static limit, 110
strangeness of particles, 54
strong equivalence principle, 62, 64
strong lensing, 82
supernova, 105, 123, 124
synchrotron radiation, 121

tachyons, 45
Taylor, Joseph, 96, 121–22
time
    absolute future/past and, *30*, 31, 38
    acceleration affecting, 17, 64–*65*, 67,
        72
    in black holes, 107
    dilation. *see* time dilation
    distance and, 17–20, 44, 100
    free fall and, 103–4
    gravity and, 64–*65*, 69, 72, 91, 101–2
    proper, 92, 93
    simultaneity of, 22–26, *23*, *24*
    slowing, 67, 101–2, 103, 104
    space-time diagrams and, 26–32, *27*,
        28, *30*
    speed of light and, 8–11, *10*, 22, 41
    speed/velocity and, *8*–11, 14–15, 17
    standing still, 11
    travel, 41, *42–43*
time dilation, 6–*13*, *7*, *8*
    atomic clocks and, 70
    formulae for, 32, 43, 70
    motion and, 67–68
    muons and, 54

in space, *18–19*
twin paradox and, 73–*75*
time-like, 37
twin paradox, 14–17, 72–78, *75*, 92–*93*

universe. *see also* space
    Big Bang of, 131, *136*, 137, 138, 139
    block universe, 38–40
    contents of, 132–33, 137
    cosmological principle of, 126, 127, 128
    curvature of, 132
    density of, 132, 133, 134–35, 137,
        138, 139
    expansion of, 129, 131, 135, *136*, 137
    flat, 127, 128, 132, 139
    inflation of, *136*, 139
    negative curvature of, 128, 132
    positive curvature of, 127–28, 131
    repulsion motion in, 136
    size of, 127, *138*
U.S. Naval Observatory, *70–71*

velocity. *see* speed of light; speed/velocity
Venus, 98
Viking Mars Lander, 98
Virgo Cluster galaxy, *119*
Voyager, *3*, 98

weak equivalence principle, 62, 64
weightlessness, 60–*61*
Wheeler, John, 106
white dwarf stars, 69, 105
white holes, 117
Wilkinson Microwave Anisotropy Probe, *136*
Wilson, Stephanie, *61*
world line, 31, 92–93
worm holes, 117

X-rays, 111, 113, *114*, 120

PICTURE CREDITS

·

Author: NASA/Source: http://www.nasa.gov/mission_pages/shuttle/shuttlemissions/sts120/multimedia/fd7/fd7_gallery.html; 83: Gravitational lensing in the galaxy cluster Abell 370 (captured by the Hubble Space Telescope).jpg/Author: NASA, ESA, the Hubble SM4 ERO Team, and ST-ECF/Source: http://www.spacetelescope.org/images/html/heic0910b.html; 94: Biosphere montreal.JPG/Author: Philipp Hienstorfer; 112–13: Einstein gyro gravity probe b.jpg/Source: NASA/User: Tano4595; 119: Phot-04a-03-hires.jpg/Author: ESO/Source: http://www.eso.org/gallery/v/ESOPIA/Galaxies/phot-04a-03-hires.jpg.html; 136: UniverseEvolution WMAP mudo.jpg/Author: Luis Fernández García/Source: NASA: Theophilus Britt Griswold – WMAP Science Team

BRIEF INSIGHTS

•

A series of concise, engrossing, and enlightening books that explore every subject under the sun with unique insight.

Available now or coming soon:

THE AMERICAN PRESIDENCY	GLOBALIZATION	PHILOSOPHY
ARCHITECTURE	HISTORY	PLATO
ATHEISM	INTERNATIONAL RELATIONS	POSTMODERNISM
THE BIBLE	JUDAISM	RENAISSANCE ART
BUDDHISM	KAFKA	RUSSIAN LITERATURE
CHRISTIANITY	LITERARY THEORY	SEXUALITY
CLASSICAL MYTHOLOGY	LOGIC	SHAKESPEARE
CLASSICS	MACHIAVELLI	SOCIAL AND CULTURAL ANTHROPOLOGY
CONSCIOUSNESS	MARX	SOCIALISM
THE CRUSADES	MATHEMATICS	STATISTICS
ECONOMICS	MODERN CHINA	THE TUDORS
EXISTENTIALISM	MUSIC	THE VOID
GALILEO	NELSON MANDELA	
GANDHI	PAUL